# Intermediate 1
# BIOLOGY

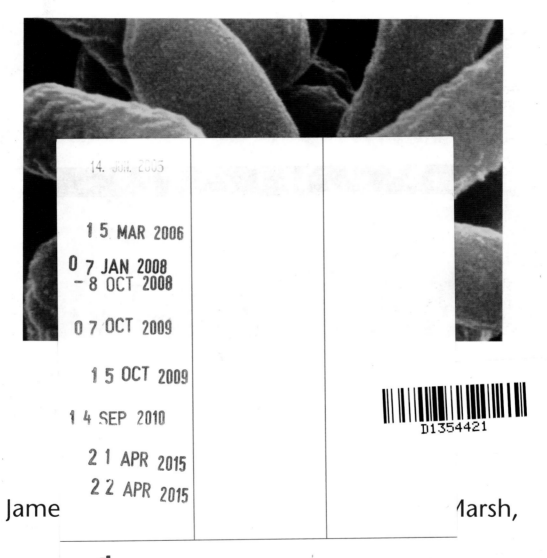

Jame                                                    Marsh,

UHI
Millennium
Institute

Hodder Gibson

## Acknowledgements

The publishers would like to thank the following for permission to reproduce material in this book. Every effort has been made to trace and acknowledge ownership of copyright. The publishers will be glad to make suitable arrangements with any copyright holders whom it has not been possible to contact.

Figure 1.1, Life File (AX/15 EJ); Figure 1.2, Mind; Figure 1.3, Life File (AJ/66A); Figure 1.5, Life File (AX/225 EJ); Figure 1.6, Life File (BA/116 A); Figure 1.7, Life File (EX/646 EJ); Figure 1.8, AJ/545 CH; Figure 2.19, Science Photo Library (P530/128); Figure 3.6, Science Photo Library (P590/211); Figure 3.23, Science Photo Library (M134/365); Figure 4.3, Science Photo Library (M260/058); Figure 4.4, Science Photo Library (M250/13); Figure 5.21, Science Photo Library (B220/772); Figure 5.22, Science Photo Library (T930/158); Figure 6.11, © Hulton-Deutsch Collection/CORBIS (HU010976); Figure 7.15, Science Photo Library (T190/316); Figure 7.19, Science Photo Library (E812/125); Figure 7.23, Science Photo Library (E820/196); Figure 8.21, Science Photo Library (B234/073); Figure 8.23, © David Reed/CORBIS (DD001588).

Figures 7.16, 7.25 and 10.24 from *The Effluent Society* by Norman Thelwell, reproduced with permission from Methuen Publishing Limited.

All other photos supplied by the authors.

Illustrations by James Torrance.

Whilst every effort has been made to check the instructions of practical work in this book, it is still the duty and legal obligation of schools to carry out their own risk assessments.

**The front cover shows a coloured scanning electron micrograph of the bacterium, *Escherichia coli* strain 0157:H7. This strain can cause food poisoning. It is found in the intestines of some cattle and can be passed to humans in milk and beef. Pasteurisation of milk and thorough cooking of meat help to prevent its spread.**

Orders: please contact Bookpoint Ltd, 130 Milton Park, Abingdon, Oxon OX14 4SB. Telephone: (44) 01235 827720. Fax: (44) 01235 400454. Lines are open from 9.00–6.00, Monday to Saturday, with a 24 hour message answering service. You can also order through our website www.hoddereducation.co.uk.

*British Library Cataloguing in Publication Data*
A catalogue record for this title is available from the British Library

ISBN-10: 0-340-81205-2
ISBN-13: 978-0-340-81205-1

Published for Hodder Gibson, 2a Christie Street, Paisley PA1 1NB.
Tel: 0141 848 1609; Fax: 0141 889 6315; Email: hoddergibson@hodder.co.uk
First Published 2004
Impression number   10 9 8 7 6 5 4 3
Year                      2010 2009 2008 2007 2006 2005

Cover photo from Science Photo Library (B230/189)
Typeset by Fakenham Photosetting Limited, Fakenham, Norfolk
Printed in Italy for Hodder Gibson, 2a Christie Street, Paisley, PA1 1NB, Scotland, UK

# Contents

Preface

# Preface

This book is intended to act as a valuable resource for pupils studying Intermediate 1 Biology during S3/4 as an alternative to Standard Grade Science and for S5 pupils studying Intermediate 1 Biology as a one-year bridging course on the way to Intermediate 2 Biology the following session. It is also suitable for use with pupils studying Access 3 Biology.

The book provides a concise set of notes that adheres to the SQA syllabus for Intermediate 1 Biology. Each section matches a unit of the syllabus; each chapter corresponds to a content area. The book contains the following special features.

- *Testing Your Knowledge*: Key questions incorporated into the text of every chapter and designed to assess continuously *Knowledge and Understanding*. These will be especially useful as homework and as instruments of diagnostic assessment to check that a full understanding of course content has been achieved.
- *Activities*: Pieces of practical work designed to give students extensive day-to-day experience of syllabus-related practical work. These will be especially useful in providing opportunities for students to gain confidence in *Practical Abilities* associated with Biology. Activities recommended for use as part of the SQA assessment are clearly indicated.
- *Applying Your Knowledge*: A variety of questions at the end of each chapter designed to give students practice in exam questions and to foster the development of specified *Problem-solving Skills* (identifying and selecting relevant information, presenting information, planning the identification and collection of data, analysing data, identifying strengths and weaknesses of experimental procedures and drawing valid conclusions). These questions will be especially useful as extensions to class work and as homework.
- *What You Should Know*: A summary of key facts and concepts at the end of each chapter in the form of a 'cloze' test and accompanied by an appropriate word bank. Each of these will provide an excellent source of material for consolidation and revision prior to the SQA examination.

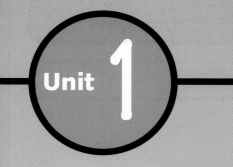

# Unit 1

# Health and Technology

**Technology is used to measure, record and monitor the state of people's health. This allows people to make informed decisions and take positive action to look after their health.**

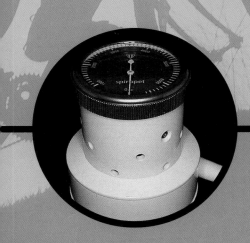

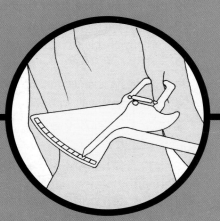

## 1 What is health and technology?

## The meaning of health

### Activity

## Identifying healthy and unhealthy aspects of lifestyle

You need ● a copy of table 1.1

| feature of person's lifestyle | | healthy | unhealthy |
|---|---|---|---|
| 1 | s/he is always able to run for a bus without getting out of breath | | |
| 2 | s/he is the correct weight for her/his height | | |
| 3 | s/he finds it easy to communicate with others and make friends | | |
| 4 | s/he smokes 20 cigarettes per day | | |
| 5 | s/he mostly feels cheerful and glad to be alive | | |
| 6 | s/he drinks only moderate quantities of alcohol | | |
| 7 | s/he is lazy and hardly ever takes any exercise | | |
| 8 | s/he often has to visit the doctor or take medicine | | |
| 9 | s/he is able to calm down easily and relax after working hard | | |
| 10 | s/he has great difficulty communicating with others and making friends | | |
| 11 | s/he is never able to run for a bus without getting out of breath | | |
| 12 | s/he has bright eyes and clear skin | | |
| 13 | s/he eats a diet rich in junk food | | |
| 14 | s/he does not smoke | | |
| 15 | s/he regularly feels down and depressed | | |
| 16 | s/he drinks alcohol to excess every weekend | | |
| 17 | s/he gets very worked up under stress and cannot cope with change | | |
| 18 | s/he hardly ever needs to visit the doctor or take medicine | | |
| 19 | s/he is far too light or far too heavy for her/his height | | |
| 20 | s/he is energetic and takes plenty of exercise | | |
| 21 | s/he is unable to calm down easily after work and stays tense | | |
| 22 | s/he has dull eyes and spotty skin | | |
| 23 | s/he eats a varied and balanced diet | | |
| 24 | s/he copes well with stress and adapts easily to change | | |

**Table 1.1** *Healthy and unhealthy aspects of lifestyle*

What to do

1 Read each feature of a person's lifestyle in the table and decide whether it indicates a healthy or an unhealthy way of life.
2 Tick the appropriate box for each statement.
3 Using the numbers, pair up each statement with its opposite number.
4 Give yourself a score out of 12 for healthy versus unhealthy features of *your* lifestyle.

2

## The health triangle

**Figure 1.1** *These teenagers are showing good physical health*

**Figure 1.2** *This charity campaigns for greater awareness of mental health problems*

Good health means much more than simply not being ill. It means enjoying a state of physical, mental and social well-being. **Physical** refers to the person's body parts such as their heart, lungs, skin, teeth and hair (see figure 1.1). **Mental** refers to the person's emotions or state of mind (see figure 1.2). If you would like more information on mental health issues, please contact the Mind website at: www.mind.org.uk. **Social** refers to the person's ability to communicate and form relationships with other people (see figure 1.3).

These three aspects of good health are often represented as the **health triangle** (see figure 1.4). If any one of the three sides is taken away, the triangle collapses. Similarly if something is wrong with one of the three aspects of a person's health, then the person suffers and is no longer healthy.

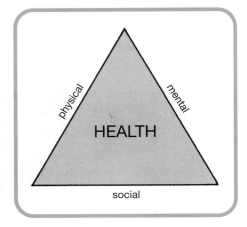

**Figure 1.4** *The health triangle*

**Figure 1.3** *Social health is important*

## Importance of a healthy lifestyle

Having a healthy lifestyle allows a person to live their life to the full since it reduces the chance of their health triangle collapsing. A healthy lifestyle includes the following:

- eating a balanced diet containing an appropriate range of foods;
- keeping fit by taking regular exercise (see figure 1.5);
- taking part in healthy, enjoyable activities (see figure 1.6);
- relaxing after working hard (see figure 1.7);
- maintaining good personal hygiene (see figure 1.8);
- avoiding unnecessary health risks such as smoking or taking drugs.

**Figure 1.5** *Regular exercise is essential for good health*

**Figure 1.6** *Cycling is a healthy, enjoyable activity*

**Figure 1.7** *Relaxing is an important part of well-being*

**Figure 1.8** *Good hygiene is essential for good health*

## Value of physiological measurements

**Physiology** is the study of the workings and functions of the human body. Physiological measurements (such as body temperature and blood pressure) are taken using special instruments. Each reading is then compared with the average value for healthy members of the population. This comparison shows whether the person's reading is in the healthy range or outside it in the unhealthy range. Physiological measurements are therefore of great value because they indicate the state of the person's health and give an unhealthy person the chance to improve their lifestyle before it is too late.

| physiological aspect of body being measured | traditional low-tech method | more advanced high-tech method |
|---|---|---|
| temperature | 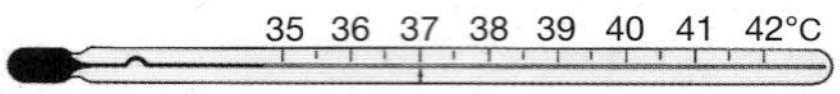 35 36 37 38 39 40 41 42°C  clinical thermometer |  20.00°C  digital thermometer |
| body fat content | 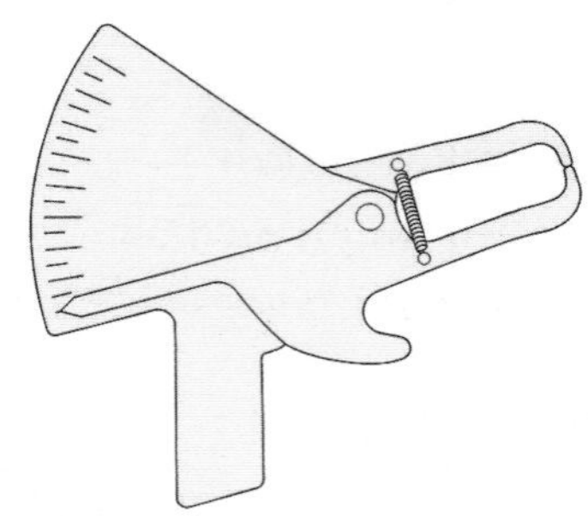 skinfold callipers | 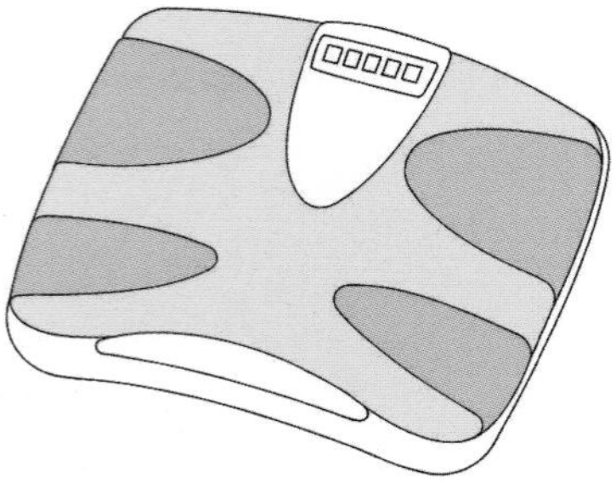 body fat sensor |
| blood pressure | 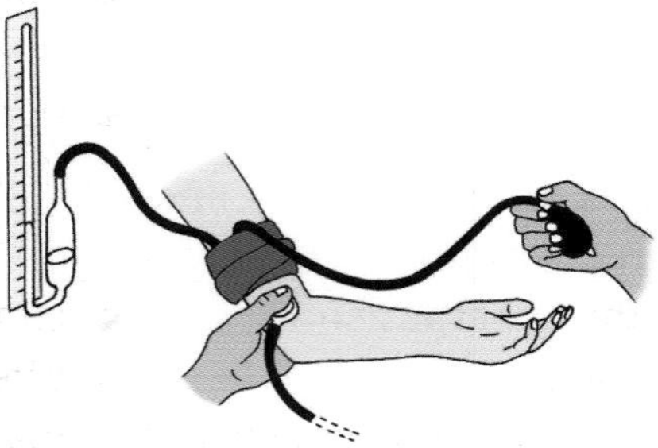 mercury manometer and stethoscope | 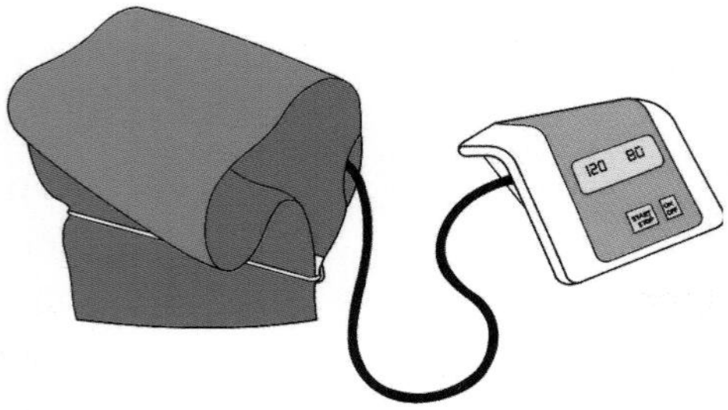 digital sphygmomanometer |
| pulse | 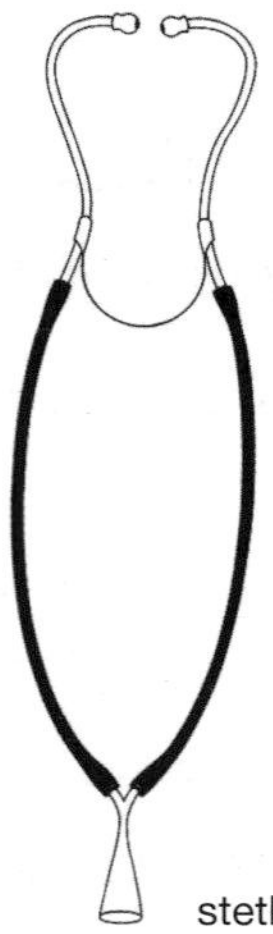 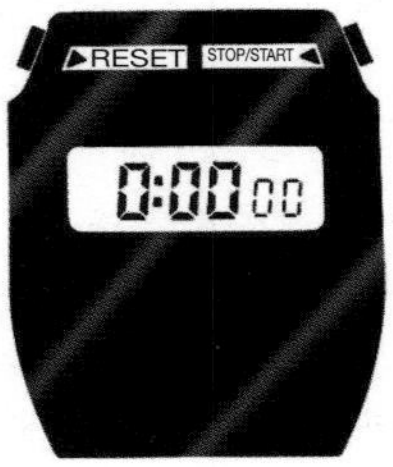 stethoscope and stopwatch | 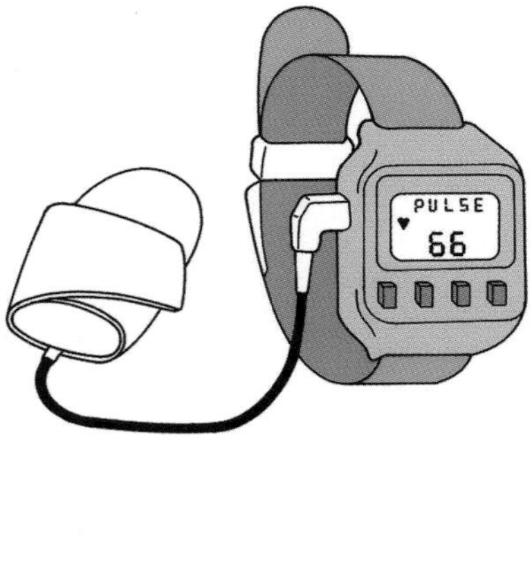 pulsometer |

**Figure 1.9** *Low-tech and high-tech instruments*

## High-tech and low-tech instruments

Two ways of taking physiological measurements are as follows:

- using **high-tech** instruments which are the products of recent advances in technology;
- using **low-tech** instruments which are the products of earlier simple technology.

Some examples of both approaches are shown in figure 1.9.

## Advantages and disadvantages

The advantages and disadvantages of high-tech and low-tech approaches to taking physiological measurements are summarised in table 1.2.

| type of instrument | advantages | disadvantages |
|---|---|---|
| low-tech | ● often cheap to make <br> ● cheap to operate | ● usually slower way to obtain results <br> ● normally cannot be connected to a computer <br> ● open to error because it depends on a human to take measurements |
| high-tech | ● usually the faster way to obtain results <br> ● can often be easily connected to a computer | ● expensive to make <br> ● often needs a high level of maintenance |

**Table 1.2**   *Advantages and disadvantages of high- and low-tech instruments*

### Testing your knowledge

1 Draw a labelled diagram of the health triangle. (3)

2 Give THREE examples of features that help to make up a healthy lifestyle. (3)

3 Give THREE examples of unnecessary health risks that should be kept to a minimum or avoided completely. (3)

4 a) Make a copy of figure 1.10 and then complete it by using arrows to connect each instrument to its use and whether it is low- or high-tech. (4)

   b) Give TWO advantages and TWO disadvantages of high-tech instruments. (4)

7

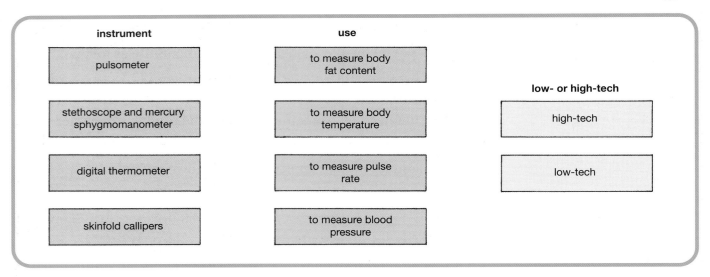

**Figure 1.10**

## Applying your knowledge

1 Copy and complete table 1.3 using examples from table 1.1. (6)

| aspect of health | example of healthy lifestyle | example of unhealthy lifestyle |
|---|---|---|
| physical | | |
| mental | | |
| social | | |

**Table 1.3**

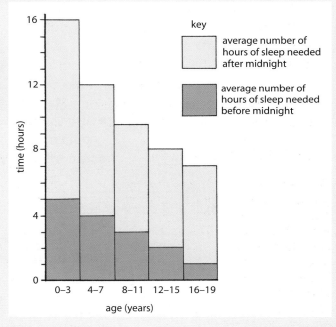

**Figure 1.11**

2 A healthy body needs rest. Sleep is the best form of rest. The bar chart in figure 1.11 shows how much sleep is needed, on average, by different age groups.

a) What is the average number of hours of sleep needed by a 15 year-old before midnight? (1)

b) What is the average number of hours of sleep needed by a 4 year-old after midnight? (1)

c) What is the total number of hours of sleep, on average, needed by a 10 year-old? (1)

d) Compared with a 13 year-old, how many more hours sleep, in total, does a 1 year-old need? (1)

e) What is the latest time at which a 14 year-old should be in bed and asleep? (1)

3 Table 1.4 shows the results of a survey involving thousands of British 15 year-olds. It compares certain aspects of the lifestyle of 15 year-olds in 1954 with those in 2004.

| aspect of lifestyle | 1954 | 2004 |
| --- | --- | --- |
| main items in diet | potatoes, bread, cereals, milk, meat, eggs, fresh fruit and vegetables | ready meals, processed food, burgers, chicken, crisps, chocolate and soft drinks |
| source of vitamin C | fresh fruit | sugary soft drinks |
| leisure activities | outdoor games and sports and limited TV viewing | computer games, Internet and much TV viewing |
| transport to school | almost always walked | only about 50% walked |
| body weight | very few teenagers were overweight and obesity was rare | about 30% were overweight and obesity was common |

**Table 1.4**

a) Amongst which group was obesity found to be common? (1)

b) The experts say that teenagers in 2004 were not necessarily eating more than in 1954. With reference to leisure activities, suggest why teenagers in 2004 were often gaining weight. (2)

c) With reference to main items in the diet, suggest why teenagers in 2004 were more likely to gain weight than those in 1954. (1)

d) Suggest why it is better to get your supply of vitamin C from a fresh orange than from a can of sugary soft drink. (1)

e) Suggest THREE ways in which an obese 15 year-old could alter their lifestyle in order to become lighter, fitter and healthier. (3)

f) Why were *thousands* of 15 year-olds included in the survey instead of just a few? (1)

**4** Staying healthy means avoiding unnecessary risks such as taking drugs. Read this passage and answer the questions.

### LEGALISE THE WEED?

Cannabis is the general name given to preparations such as marijuana and hashish made from the Indian hemp plant (see figure 1.12). When cannabis is smoked, the effect begins within a few minutes and can last for hours. Users feel high, lose their inhibitions and become very talkative yet calm. When the drug wears off, they feel lethargic, hungry and sleepy. Regular users sometimes suffer anxiety attacks and gaps in their memory.

Some people argue that use of cannabis should be made legal since it is less harmful than alcohol or tobacco. However, evidence suggests that marijuana causes even more damage to the lungs than tobacco smoke. A 'joint' a day is equivalent to smoking a full pack of cigarettes. This is because the joint contains 6 mg of tar compared with 1.2 mg in a cigarette. In addition, cannabis smokers inhale more deeply and hold the smoke in their lungs for longer.

Some people argue that our society already has enough problems with legal drugs (alcohol and tobacco). Legalising cannabis, they claim, would not give people a less harmful alternative. Instead it would provide an *additional* drug which is harmful to people's health.

Can smoking this seriously damage your health?

Cannabis sativa

**Figure 1.12**

a) (i) Name TWO forms of cannabis.

(ii) What is the source of these? (2)

b) Why do people smoke cannabis? (1)

c) How does a user normally feel when the effect of the drug has worn off? (2)

d) (i) By how many times is the tar content of a 'joint' greater than that of a cigarette?

(ii) What behaviour on the part of the user makes a joint more damaging to their health than a cigarette? (2)

e) Some people say that making cannabis legal would provide a harmless alternative to alcohol and tobacco. What argument against this viewpoint is given in the passage? (1)

5 In table 1.5, each ★ represents one **high-risk** point. High-risk points are linked to certain aspects of an unhealthy lifestyle which make the person more likely (on average) to suffer a major illness.

| aspect of un-healthy lifestyle | major illness | | | |
|---|---|---|---|---|
| | **heart attack** | **liver disease** | **lung cancer** | **stroke** |
| eats a diet rich in butter and fried food | ★★★ | | | ★ |
| drinks alcohol to excess | | ★★★ | | |
| smokes heavily | ★★★ | | ★★★ | ★★ |
| takes a lot of salt on food | ★ | | | ★★ |

**Table 1.5**

a) Which aspect of an unhealthy lifestyle causes liver disease? (1)

b) How many 'heart attack' high-risk points are caused by eating a diet rich in butter and fried food? (1)

c) How many high-risk points in total are caused by taking a lot of salt on your food? (1)

d) Which aspect of an unhealthy lifestyle is linked to the greatest number of high-risk points? (1)

e) Which TWO major illnesses are linked to high-risk points from three different aspects of an unhealthy lifestyle? (2)

6 Stress can affect a person's health in many ways. Some of these are given in the list below.

anxiety          inability to cope

constipation       inability to show feelings

depression        indigestion

diarrhoea         itchy skin

headache         short temper

a) Draw a table and group these effects under the headings **physical effects** and **mental effects**. (4)

b) What advice would you give to a friend whose job is so stressful that they are suffering from many of these effects? (1)

7 An important way to stay healthy is to take steps to avoid illness. One way of protecting the body against certain diseases is to be vaccinated at an early age. The vaccine then prepares the body in advance. If the germs get into the body, its defences are already in place. Table 1.6 shows a plan of vaccinations recommended for children and teenagers in the UK.

a) How many different diseases are given in the table? (1)

b) (i) At what age should the young person be given their first vaccination against measles according to this plan?

(ii) How many vaccinations should the person get against measles? (2)

c) (i) At what age should the person be given their first dose of vaccine against polio and tetanus?

| age | vaccine | notes |
|---|---|---|
| during first year of life | polio, tetanus, diphtheria, whooping cough and meningitis | 1st dose at 2 months, 2nd dose at 3 months, 3rd dose at 4 months |
| during second year of life | measles, mumps and rubella, or measles only | single dose |
| 4–5 years | polio, tetanus and diphtheria | booster dose |
| 10–14 years | tuberculosis (BCG) and rubella | single dose of each |
| 15–19 years | polio, tetanus and diphtheria | booster dose |

**Table 1.6**

(ii) How many doses of vaccine against these two diseases are recommended during the first 19 years of life in this plan? (2)

d) Name TWO vaccines that are recommended for use with school pupils aged 10–14 years. (2)

8 Healthy people are fit. Fitness is a combination of the three 'S' factors: stamina, suppleness and strength. When carried out regularly, some activities improve all three 'S' factors, while others only affect one or two of them as shown in table 1.7.

Match the activities in the table with the numbers in figure 1.13. Number 2, for example, increases stamina and suppleness but does not affect strength, so it is disco dancing. (6)

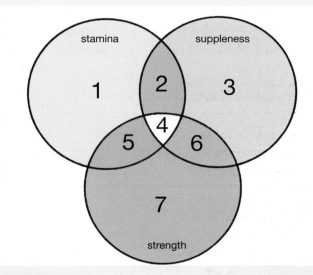

**Figure 1.13**

| activity | significant increase in stamina? | significant increase in suppleness? | significant increase in strength? |
|---|---|---|---|
| climbing stairs | yes | no | yes |
| disco dancing | yes | yes | no |
| sailing | no | yes | yes |
| swimming | yes | yes | yes |
| walking quickly | yes | no | no |
| weight lifting | no | no | yes |
| yoga | no | yes | no |

**Table 1.7**

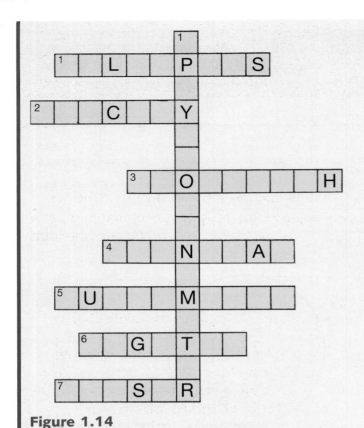

**Figure 1.14**

**9** Solve the puzzle in figure 1.14 using the following clues:

*Down*

1 High-tech instrument used to measure blood pressure.

*Across*

1 Low-tech instrument used to measure body fat.
2 Silvery liquid found in low-tech instrument for measuring blood pressure.
3 Low-tech instrument used along with finger to measure pulse rate.
4 Low-tech type of thermometer.
5 High-tech instrument for measuring pulse rate.
6 High-tech type of thermometer.
7 High-tech instrument for measuring body fat.

(8)

## What you should know

| | | |
|---|---|---|
| calliper | lifestyle | relaxing |
| computer | low | risks |
| drugs | measurements | social |
| expensive | physical | triangle |
| high | pulsometer | varied |

**Table 1.8** *Word bank for chapter 1*

**1** Good health means more than not being ill. It means being in a state of _____, mental and _____ well-being.

**2** These three aspects of well-being make up the health _____.

**3** To stay healthy, a person needs to adopt a healthy _____.

**4** A healthy lifestyle includes eating a good _____ diet, exercising, enjoying pleasurable activities, _____ and avoiding health _____ such as taking _____.

**5** Physiological _____ indicate the state of a person's health.

**6** These measurements can be taken by using _____ -tech instruments such as a digital sphygmomanometer or a _____ or by using low-tech instruments such as a skinfold _____ or a stopwatch.

**7** High-tech instruments usually give results quickly but are _____ to produce. _____ -tech instruments are cheap to make but normally cannot be connected to a _____.

# 2 A healthy heart

## Circulatory system

The **circulatory system** in the human body is made up of the **heart** and the **blood vessels**. The heart is a muscular organ that pumps blood around the body. The blood vessels are a system of tubes that carry blood to all parts of the body.

### Activity

#### Examining a sheep's heart

**Information**

This activity is intended as a teacher demonstration.

**Teacher needs**
- sheep's heart
- dissecting board
- dissection instruments
- pair of disposable gloves

**Pupils may need**
- pair of disposable gloves

**What to do**

1 Put on disposable gloves if you intend to handle the sheep's heart.
2 Hold the heart with both hands and feel its muscular wall.
3 Hold the heart in one hand and compare its size to that of a clenched fist.
4 Return the heart to the teacher for dissection. Figure 2.1 shows a sheep's heart before dissection and after the front half has been cut away. The heart's four chambers are numbered 1–4 on the diagram.

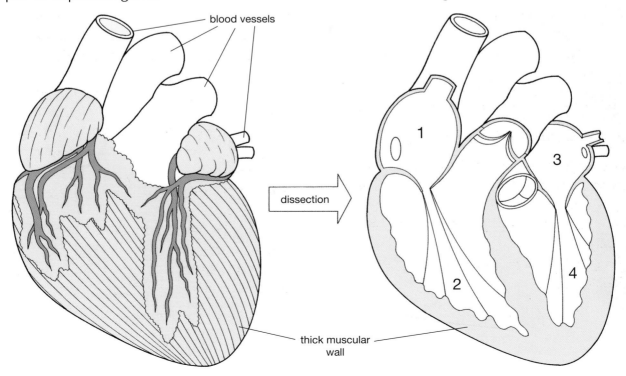

blood vessels

dissection

thick muscular wall

**Figure 2.1**  *Structure of a sheep's heart*

## Heart

Figure 2.2 shows a simplified version of the inside of a human heart. The top two chambers (1 and 3) collect blood and pass it down to the lower chambers (2 and 4). During each heart beat, the muscle in the heart wall contracts and pumps blood out of these chambers at the same time. Blood in chamber 2 is sent to the lungs. Blood in chamber 4 is sent to all parts of the body.

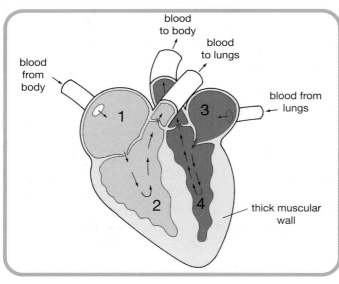

**Figure 2.2**   *Section through a human heart*

### Valves

There are four valves inside the heart. For example, one is found between chambers 1 and 2 and another between chambers 3 and 4. Each valve opens to allow blood to pass through in the correct direction from one place to another. The valve then closes to prevent blood flowing backwards in the wrong direction.

The sounds of these valves closing as the heart beats can be heard using a low-tech instrument called a stethoscope (see figure 2.3). Doctors often use a stethoscope to listen to a patient's heart beating and check that the valves are working normally.

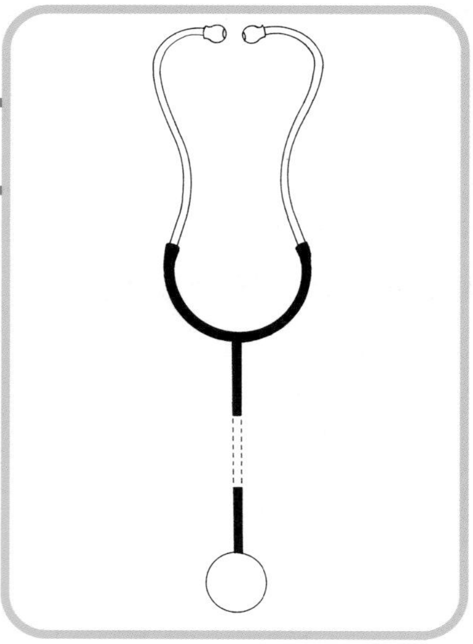

## Blood vessels

The route taken by blood as it flows through the heart and around the body in blood vessels is shown in figure 2.4. This diagram also shows details of the three main types of blood vessel – arteries, veins and capillaries.

### Artery

**Figure 2.3**   *A stethoscope*

An artery is a large blood vessel that carries blood *away from* the heart. It has a thick muscular wall because it carries blood at high pressure from the heart.

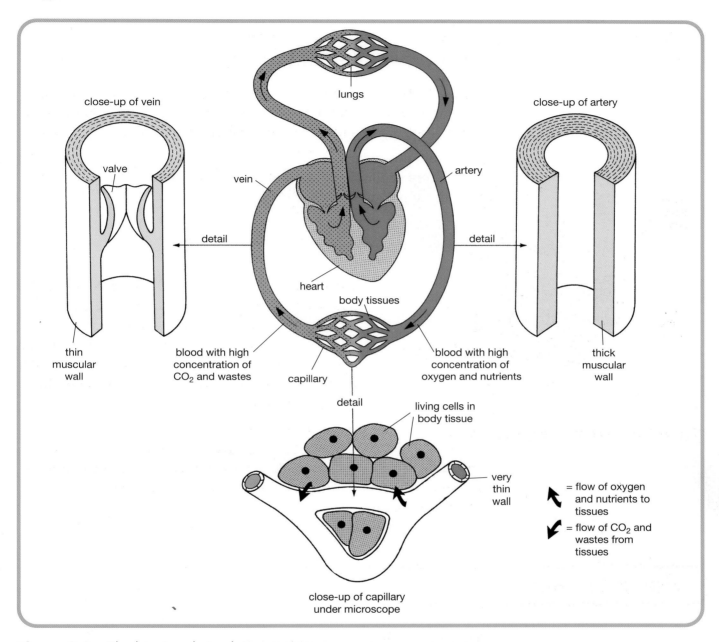

**Figure 2.4** *The heart and circulatory system*

## Capillary

Each artery divides and subdivides into smaller and smaller vessels. Eventually the vessels are microscopic and their walls are very thin. These tiny vessels are called **capillaries** and they are in close contact with all living cells and tissues. (A tissue is a group of similar cells that work together.) Because their walls are so thin, capillaries allow the **oxygen** and **nutrients** that they are carrying to pass out of the blood into the surrounding tissues. At the same time, **carbon dioxide** and other **wastes** leave the tissues and pass through the thin capillary walls into the blood to be carried away and removed from the body.

Tiny capillaries unite into larger vessels which join up into still larger vessels until these in turn make up veins. Capillaries are therefore described as the link between arteries and veins.

## Vein

A **vein** is a large blood vessel that carries blood *towards* the heart. It has a thin muscular wall because the blood passing through it has come from capillaries and is now at low pressure. Valves are present in veins to prevent backflow of blood.

### Activity

## Demonstrating the emptying of part of a vein in the arm

### Information
Up until the 1600s, people thought that blood ebbed and flowed inside the body like the tide coming in and going out. However in the early 1600s, a doctor called William Harvey had a different idea. He felt sure that blood travelled round the body in a circle. In his famous experiment he demonstrated that blood in veins cannot travel backwards but can only go in one direction from the tissues towards the heart.

### You need
- scarf
- brush handle or window pole

### What to do (also see figure 2.5)
1 To repeat part of Harvey's experiment, decide who is to be the subject and who is to carry out steps 2–9 below as the operator.
2 Ask the subject to grasp the brush handle tightly.
3 Tie a scarf tightly round the subject's bare arm just above the elbow. Look for arm veins which will now become obvious (especially in boys).
4 Place fingers A and B on a large vein as shown in the diagram.
5 Press finger A down hard.
6 Slide finger B along the vein away from finger A several times.
7 Remove finger B and observe a swelling in the vein showing the position of a closed valve.
8 Look carefully at the length of vein between finger A and the closed valve and decide if it is empty or full of blood.
9 Remove finger A and decide if the same length of vein is empty or full of blood after a few seconds.
10 Copy and complete the following paragraph using the words in the word bank.

A portion of a vein was emptied using finger B. While finger A was being pressed down hard on the _____, the portion of vein between finger A and the closed valve stayed _____. This showed that the _____ was preventing blood from flowing _____ and that blood can only flow in one _____. Once finger A was lifted, blood flowed again from the _____ along the vein towards the _____. The portion of vein that had been empty soon _____ up again. This demonstration supports Harvey's idea that _____ flows round the body in a _____ and not back and forth like tides.

(**Word bank**: backwards, blood, circle, direction, empty, filled, hand, heart, valve, vein)

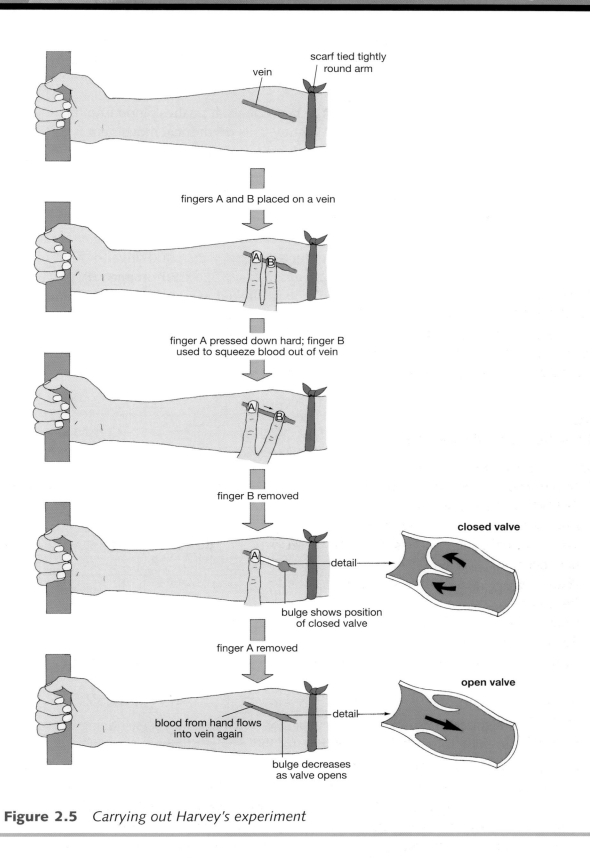

**Figure 2.5** *Carrying out Harvey's experiment*

## Pulse

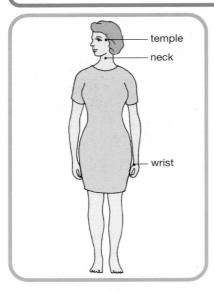

temple
neck
wrist

Each time the heart beats, it pushes blood into the arteries making them swell slightly. This rhythmical movement is called a **pulse**. It can be felt by pressing a finger tip on an artery that is near the skin surface. The pulse can be felt at different positions in the body. Three examples are shown in figure 2.6.

**Pulse rate** is a measure of the number of times that the heart beats per minute. Pulse rate can be measured using either low-tech instruments such as a **stopwatch** and finger or high-tech instruments such as a **pulsometer** (see figure 2.7 on page 19) or a heart rate **monitor** linked to a computer (see figure 2.9 on page 21).

**Figure 2.6**   *Three pulse points*

### Activity

## Measuring pulse rate using a stopwatch (assessment)

You need
- stopwatch
- calculator

What to do
1 Make a copy of table 2.1.
2 Using your stopwatch, time yourself sitting quietly for two minutes to allow your pulse to settle at a steady rate.

| position of body | pulse rate (beats/15 s) | | | average pulse rate (beats/15 s) | average pulse rate (beats/min) |
|---|---|---|---|---|---|
| | trial 1 | trial 2 | trial 3 | | |
| sitting | | | | | |
| standing up | | | | | |

**Table 2.1**   *Pulse rate results*

3 Locate your pulse. (You will probably find that the pulse point in your neck is the easiest to locate and use while operating the stopwatch with your other hand.)
4 Measure your pulse rate for 15 seconds while sitting quietly. This is called your resting pulse rate. Enter the result in your table.
5 Repeat step 4 twice.
6 Stand up straight and unsupported. Use your stopwatch to time yourself doing this for two minutes to allow your pulse to settle at a steady rate for this new position.
7 Measure your pulse rate for 15 seconds while standing up straight and unsupported. Enter the result in your table.
8 Repeat step 7 twice.
9 Complete the table by calculating average values. (*Note*: Taking several readings and calculating an average gives an overall result that is *more reliable*.)
10 a) State which body position gives the higher pulse rate.
   b) Suggest a reason for your result.
11 Ask the teacher to assess your work.

## Measuring pulse rate using a pulsometer

**Figure 2.7**  *A pulsometer*

**You need**
- pulsometer (see figure 2.7)

**What to do**
Follow the procedure shown in figure 2.8.

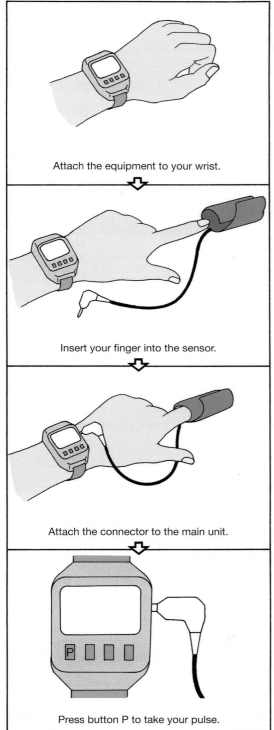

Attach the equipment to your wrist.

Insert your finger into the sensor.

Attach the connector to the main unit.

Press button P to take your pulse.

**Figure 2.8**  *Using a pulsometer*

### Normal pulse

Pulse rate is expressed as beats per minute and gives a direct measure of the rate at which the heart is beating. Pulse rate varies considerably from person to person and within the same person under varying circumstances.

### Resting pulse rate

This is a measure of pulse rate when the body is at rest and has not been exercising for at least a few minutes. On average, resting pulse rate for men is about 78 beats/minute and for women about 84 beats/minute, though any value between 50 and 100 is regarded by experts as being within the normal range. Many factors affect pulse rate. Some examples are shown in table 2.2.

| factor | effect on pulse rate |
|--------|---------------------|
| body size | increases as person becomes heavier |
| age | lower in adults than in children |
| sex | slightly higher in women than in men |
| fitness | decreases as person becomes fitter |

**Table 2.2**  *Factors affecting pulse rate*

### Pulse as a health indicator

If a person is fit, their heart muscle is bigger and more efficient than that of an unfit person. Their heart muscle needs to contract less often to pump blood round the body compared with the heart muscle of an unfit person. So the fit person tends to have a lower pulse rate.

In general the lower the resting pulse rate, the fitter the person. However it must be kept in mind that much variation exists between people and there are exceptions to the rule.

A very high resting pulse rate (e.g. above 100 beats/minute) is normally unhealthy and can lead eventually to heart disease.

## Recovery time

Recovery time is the time taken for pulse rate to return to its normal resting value after exercise. The shorter the recovery time, the fitter the person.

### Activity

## Investigating the effect of exercise on pulse rate

*You need*
- stopwatch
- calculator

*What to do*

1 Make a copy of table 2.3.
2 Using your stopwatch, time yourself sitting quietly for two minutes to allow your pulse to settle at a steady rate.
3 Locate your pulse.
4 Measure your pulse rate for 30 seconds while continuing to sit quietly.

| when heart rate was recorded | pulse rate | |
|---|---|---|
| | beats/30 s | beats/min |
| before exercise | | |
| immediately after exercise | | |
| 1 min after exercise | | |
| 2 min after exercise | | |
| 3 min after exercise | | |
| 4 min after exercise | | |
| 5 min after exercise | | |
| 6 min after exercise | | |

**Table 2.3**  *Results of effect of exercise on pulse rate*

5 Repeat step 4 twice and calculate your average resting pulse rate in beats/30 s.
   (*Note*: Taking several readings and calculating an average gives an overall result that is *more reliable*.)
6 Enter the result in the table for 'before exercise'.
7 Exercise vigorously by running on the spot for two minutes.
8 Stop exercising and immediately take your pulse for 30 seconds.
9 Record the result in the table.
10 Take your pulse at one minute after you stopped exercising and record the result as before.
11 Continue to take and record your pulse every minute up to 6 minutes from when you stopped exercising.
12 Complete the table by calculating your pulse in beats per minute for each reading.
13 Answer the following questions
   a) What effect did exercise have on your pulse rate?
   b) How long was your recovery time?
   c) How many pupils in the class had a shorter recovery time than you and therefore were fitter than you?
   d) How many pupils in the class had a longer recovery time than you and therefore were less fit than you?

## Activity

### Investigating the effect of exercise on pulse rate using a heart monitor linked to a computer

**You need**

- stopwatch
- heart rate monitor (e.g. Pasco heart rate sensor attached to a computer)

**What to do**

1 Prepare a copy of table 2.4.

2 Using the stopwatch, time yourself sitting quietly for two minutes to allow your pulse to settle at a steady rate.

3 Attach the heart rate sensor to your earlobe or to the web of skin on your hand between the thumb and forefinger (see figure 2.9).

4 Watch the screen as the computer records your pulse rate automatically. After one minute enter the reading in your table for 'before exercise'. (This is your resting pulse rate in beats/minute.)

5 Unclip the sensor and exercise vigorously for two minutes.

6 Attach the sensor again and take a pulse rate reading as soon as possible and then every minute for 6 minutes. Enter each result in the table as you go along.

7 Answer the following questions.

  a) What effect did exercise have on your pulse rate?

  b) How long was your recovery time?

  c) (i) Who was the fittest of all the pupils who tried this activity?

    (ii) How could you tell?

| when heart rate was recorded | pulse rate (beats/min) |
|---|---|
| before exercise | |
| immediately after exercise | |
| 1 minute after exercise | |
| 2 minutes after exercise | |
| 3 minutes after exercise | |
| 4 minutes after exercise | |
| 5 minutes after exercise | |
| 6 minutes after exercise | |

**Table 2.4** *Pulse rate results from a computer*

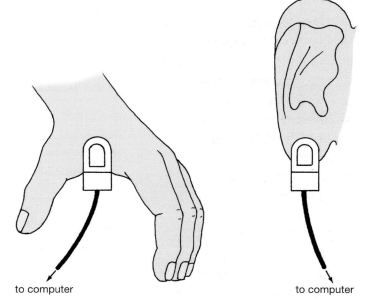

to computer           to computer

**Figure 2.9** *Two ways of attaching a heart rate sensor*

## Reducing pulse rate and recovery time

In general, fit people have low resting pulse rates and short recovery times. Their heart and lungs are so healthy and efficient that their pulse returns quickly to its normal low rate. The reverse is true of unfit people.

Pulse rate and recovery time can be reduced by taking more exercise and becoming fitter. This does not mean engaging in sudden bursts of activity and then lazing around the rest of the time. It means being an active person in general and adopting good lifestyle habits such as:

- playing a sport that you enjoy on a regular basis;
- finding a form of exercise that you like doing and doing it for at least 20 minutes, three times every week;
- climbing the stairs instead of taking the lift;
- sometimes walking or cycling to school or work instead of always taking the bus or getting a lift.

## Testing your knowledge

**1** a) What job does the heart do? (1)
b) Why are valves present in the heart? (1)

**2** Copy and complete figure 2.10 by using arrows to connect the type of blood vessel with its function and the description of its wall. (6)

**3** a) Name blood vessels X and Z in figure 2.11. (2)
b) Name a substance that passes from living cells into the blood at arrow Y. (1)

**4** a) What causes pulse? (1)
b) Name THREE places in the body where pulse can easily be felt using a finger. (3)

**5** a) What is meant by the term *recovery time*? (2)
b) If a person has a short recovery time, what does this indicate about their level of fitness? (1)
c) What can a person do to reduce their recovery time? (2)

| blood vessel | function | description of wall |
|---|---|---|
| artery | carries blood towards the heart | very thin and non muscular |
| capillary | carries blood away from the heart | thick and muscular |
| vein | allows nutrients and oxygen to pass from blood to tissues | thin and muscular |

**Figure 2.10**

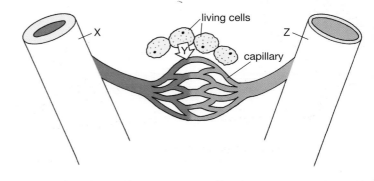

**Figure 2.11**

## Blood pressure

When the heart muscle contracts, it pumps blood out into the arteries. The blood in arteries is therefore under pressure. This pressure is called blood pressure. It is at its highest level just after the heart muscle has contracted. It is at its lowest level in between contractions when the heart is filling up again.

## Measuring blood pressure

Blood pressure can be measured using a stethoscope and a mercury manometer. This is the low-tech approach as shown in figure 2.12.

Figures 2.13 and 2.14 show high-tech methods of measuring blood pressure using a digital sphygmomanometer. The apparatus in figure 2.13 has a cuff that has to be inflated manually; the apparatus in figure 2.14 is more advanced and has a cuff that inflates automatically. In each case, the instrument is used to take two readings. These are measured in millimetres of mercury (mm Hg).

- The first (**higher pressure**) occurs when the heart beats and pumps blood into the arteries. The average reading is 120 mm Hg.
- The second (**lower pressure**) occurs between beats when the heart relaxes and fills with blood. The average reading is 80 mm Hg.
- These two values are often written in the following brief way: 120/80.

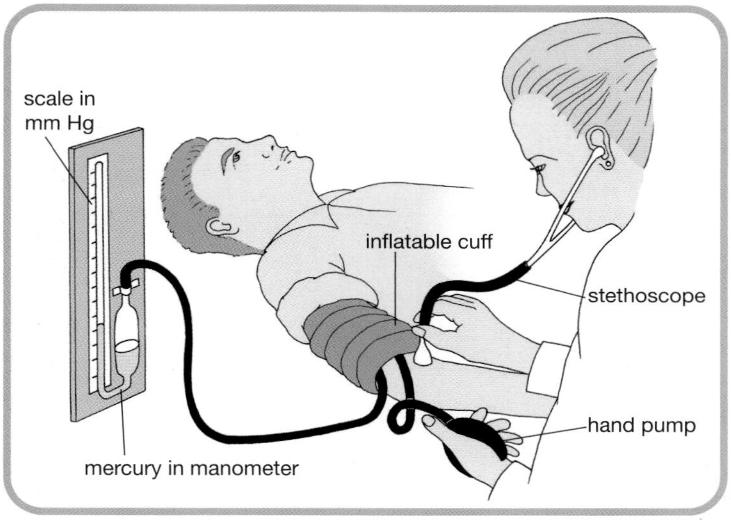

**Figure 2.12** *A low-tech method of measuring blood pressure*

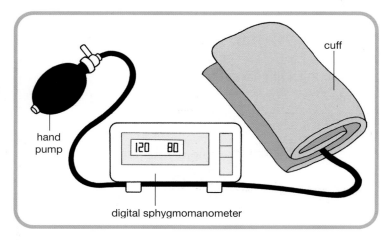

**Figure 2.13** *A digital sphygmomanometer with a manually inflated cuff*

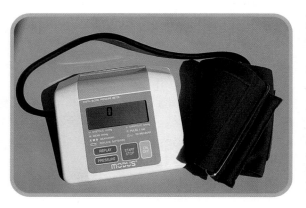

**Figure 2.14** *A digital sphygmomanometer with an automatically inflated cuff*

## Activity

# Measuring blood pressure using a digital sphygmomanometer

### Information

This activity is intended as a demonstration to be carried out by the teacher.

### Teacher needs

● digital sphygmomanometer

| person | state of person | first higher blood pressure (mm Hg) | second lower blood pressure (mm Hg) | brief way of showing the two results |
|--------|-----------------|-------------------------------------|-------------------------------------|--------------------------------------|
| 1 | sitting | | | |
| | after exercise | | | |
| 2 | sitting | | | |
| | after exercise | | | |
| 3 | sitting | | | |
| | after exercise | | | |

**Table 2.5**   *Blood pressure results*

### What to do

1 Prepare a copy of table 2.5.
2 Watch the demonstration as the teacher:
  ● turns on the digital sphygmomanometer's power;
  ● presses the valve to deflate the cuff if necessary;
  ● asks volunteer 1 to be seated and wraps the cuff around their bare upper left arm;
  ● ensures that the cuff is clear of the elbow and that it is loose enough to allow two fingers to be placed beneath its edge;
  ● presses the 'constant air release' side of the valve;
  ● inflates the cuff until beeping sounds are heard;
  ● stops inflating to allow the pressure to decrease;
3 Look at the 'high' and 'low' blood pressure readings on the display (and the pulse reading) .
4 Enter the 'sitting' results for person 1 in the table.
5 Watch the demonstration as the teacher carries out the following steps:
  ● removes the cuff from person 1's arm;
  ● gets person 1 to exercise for three minutes and repeats the above procedure.
6 Enter the 'after exercise' results for person 1 in the table.
7 Watch the demonstration as the teacher repeats the procedure with two more volunteers and enter the results in the table.
8 Draw a conclusion about the effect of exercise on blood pressure from the results.

# 'Normal' blood pressure

The chart in figure 2.15 gives an idea of which blood pressure values are regarded as normal, high or low. This chart is fairly *reliable* because it is based on *average* results calculated using data from *many thousands* of people. However, it must be kept in mind that blood pressure is affected by many factors (see below) and varies widely depending upon circumstances. What is 'normal' for one person is not necessarily 'normal' for someone else.

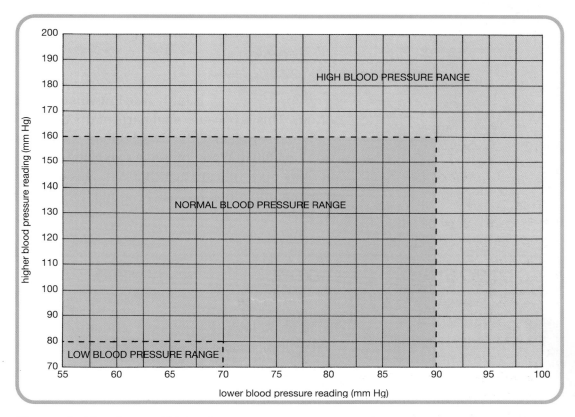

**Figure 2.15**  *Range of blood pressure*

## High blood pressure

**High blood pressure** is rare in young people. It is fairly common in adults over the age of 35 and can be caused by:

- being overweight;
- not taking enough exercise;
- eating a diet containing too much fatty food (especially animal fat);
- eating a diet containing too much salt;
- drinking alcohol excessively;
- having a stressful lifestyle.

### Effect on health

A person who has high blood pressure does not feel any different from normal. The only way of finding out if your blood pressure is high is to have it checked. If high blood pressure goes undetected for a long time, it can eventually lead to serious health problems. Some of these are as follows:

- **angina** – intense cramp-like pain across the chest indicating that the heart muscle is not getting enough blood;
- **heart attack** – damage to the heart muscle resulting from blockage of a blood vessel and lack of oxygen;
- **stroke** – blockage or bursting of a blood vessel in the brain that may cause brain damage.

### Low blood pressure

**Low blood pressure** is fairly rare but when it does occur, it can be serious. This is because vital parts of the body such as the brain are not being supplied with enough blood carrying oxygen and nutrients. A sudden fall in blood pressure can lead to fainting. Low blood pressure may indicate heart failure. This occurs when the heart is not able to pump blood out hard enough into the arteries.

## Composition of blood

Blood is made up of red blood cells and white blood cells in a watery yellow liquid called plasma. Plasma contains dissolved substances such as sugar. Figure 2.16 shows the proportions of the parts that make up a blood sample. Figure 2.17 shows blood highly magnified under a microscope.

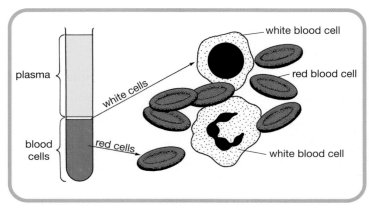

**Figure 2.16**  *Composition of blood*

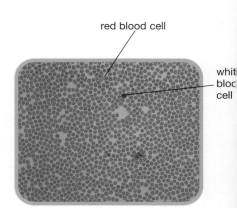

**Figure 2.17**  *Blood cells as seen under the microscope*

## Red blood cells

These are tiny disc-shaped cells. They are very numerous. One cubic millimetre ($mm^3$) of blood contains approximately 5 million red blood cells. Their job is to transport oxygen from the lungs to the living tissues.

## White blood cells

Compared with red blood cells, white blood cells are less numerous. One cubic millimetre ($mm^3$) of blood contains 5000–9000 white blood cells. Their job is to protect the body against invasion by micro-organisms.

# Blood tests and cell counts

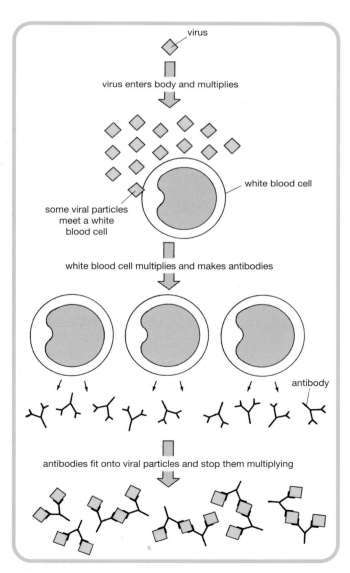

## Detection of infection

When a virus attacks and infects the body, the body tries to defend itself. White blood cells are the body's main line of defence. They respond to the presence of a virus by making antibodies. These protective molecules fit onto the virus particles and stop them from multiplying further and causing disease (see figure 2.18).

By testing a person's blood for the presence of antibodies, doctors can tell whether or not a person is infected with a particular virus even if they are not showing any symptoms of the viral disease. Viral infections, including hepatitis (B and C) and HIV (human immunodeficiency virus) which causes AIDS, can be detected in this way.

**Figure 2.18** *Action of antibodies*

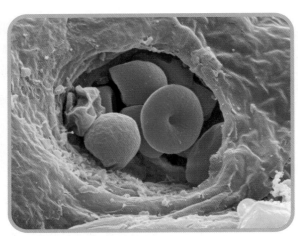

**Figure 2.19** *Red blood cells in a capillary*

## Detection of anaemia

Oxygen is transported round the body from the lungs to the tissues by red blood cells (see figure 2.19). These cells are able to do this job because they contain a chemical called **haemoglobin**. Haemoglobin picks up oxygen at the lungs and releases it in the tissues. Haemoglobin is rich in **iron** and red in colour.

**Anaemia** is a blood disorder in which the number of red blood cells and/or haemoglobin concentration is abnormally low. The most common cause of this condition is **deficiency** in iron. By testing a person's blood for iron content, doctors can detect whether or not the person is suffering from anaemia.

## Detection of diabetes

Glucose sugar is transported round the body to living tissues in blood plasma. The concentration of glucose in blood plasma is normally kept at the correct level by a hormone (chemical messenger) called **insulin**. If a person does not make enough insulin, their blood sugar level goes up. This disorder is called **diabetes**. By testing a person's blood for sugar concentration, doctors can detect whether or not the person is suffering from diabetes and give appropriate treatment. Figure 2.20 shows a pen used to inject insulin.

## Detection of leukaemia

In normal blood, there are several hundred red blood cells to every one white blood cell (see figure 2.21). New blood cells are made in the bone marrow. If this tissue develops a form of cancer, it makes too many white blood cells. This results in **leukaemia** which is a disorder of the blood where the number of white blood cells is much higher than normal.

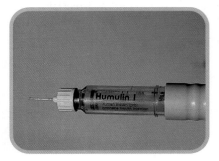

**Figure 2.20** *A pen for injecting insulin*

Doctors can examine a thin smear of blood under a microscope and carry out a **white blood cell count**. If it is abnormally high (see figure 2.22), this indicates leukaemia. Sometimes a bone marrow transplant is carried out (especially in young children) to treat leukaemia.

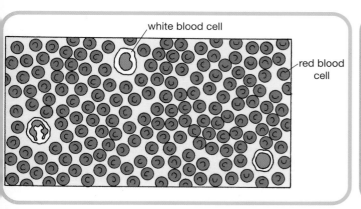

**Figure 2.21**  *The composition of normal blood*

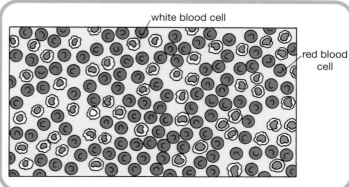

**Figure 2.22**  *Blood from a sufferer of leukaemia*

## Identification of blood groups

A person's blood belongs to one of four possible groups called A, B, AB and O. It also belongs to one of two further blood group factors called **Rhesus positive (Rh+)** and **Rhesus negative (Rh−)**. Tests are used to identify a person's blood group. For example, one person could be blood group A, Rh−; another person could be blood group O, Rh+ and so on.

A person who needs a transfusion of blood cannot simply receive blood from any other person who is willing to donate blood. Some blood groups will not mix freely with others. Instead the red blood cells stick together as **clumps** and block small blood vessels. This leads to circulation problems and serious or even fatal damage to tissues.

Table 2.6 shows which groups can safely receive blood from which donors and which cannot. For example, people with blood group A can receive blood from people with blood group A or O but not from groups B or AB. Table 2.7 shows in a simple way how Rh+ blood must not be given to a Rh− person.

| | | group donating blood | | | |
|---|---|---|---|---|---|
| | | **A** | **B** | **AB** | **O** |
| group **receiving** blood | A | ✓ | ✗ | ✗ | ✓ |
| | B | ✗ | ✓ | ✗ | ✓ |
| | AB | ✓ | ✓ | ✓ | ✓ |
| | O | ✗ | ✗ | ✗ | ✓ |

key ✓ = successful transfusion
✗ = unsuccessful transfusion

**Table 2.6** *Successful and unsuccessful transfusions*

| | | group donating blood | |
|---|---|---|---|
| | | **Rh+** | **Rh−** |
| group **receiving** blood | Rh+ | ✓ | ✓ |
| | Rh− | ✗ | ✓ |

key ✓ = successful transfusion
✗ = unsuccessful transfusion

**Table 2.7** *Rhesus blood factor transfusions*

## Occurrence of blood groups

The occurrence of blood groups varies from one ethnic group to another. A few examples are given in table 2.8.

| people (ethnic group) | A | B | AB | O |
|---|---|---|---|---|
| British | 35 | 16 | 4 | 45 |
| North American Indians | 7 | 2 | 0 | 91 |
| Chinese | 31 | 28 | 7 | 34 |
| Maoris | 56 | 3 | 1 | 40 |

**Table 2.8** *Occurrence of blood groups*

## Testing for alcohol and drugs

Many people enjoy alcoholic drinks because the alcohol has a calming effect and makes them feel relaxed and at ease. If alcohol is drunk in moderation, it is probably harmless. However alcohol is the most commonly abused drug in Britain. A large percentage of the population drink above sensible limits (see figure 2.23). Many of these people are socially disruptive for the following reasons:

- they suffer severe mood swings involving fits of depression and bouts of violence;

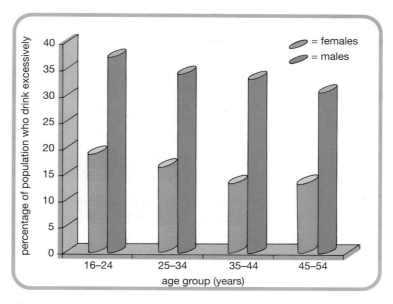

**Figure 2.23**

- they cannot manage their family life so their partner and children suffer;
- they cannot do their job properly.

A person who is under the influence of alcohol at work is unable to concentrate on the job or control their body properly. They are a safety risk because they are more likely to have an accident. This could lead to themselves or other people being injured or property being damaged.

On the roads, drunk drivers are responsible for about 1000 deaths per year in Britain. The most common cause of death amongst people under the age of 25 is a drink-related road accident. The legal limit for drivers is 80 mg alcohol/100 cm³ blood. The law allows the police to test suspected drunk drivers using breath alcohol equipment such as breathalysers and, more commonly, alcometers (see page 88). A failed roadside breath alcohol test is followed by a blood or urine test at the police station to measure blood alcohol concentration more accurately.

In 2003 police began trials on a new saliva test specially designed to use at the roadside with suspected drug drivers. It is likely that in the future, testing for alcohol and drugs in the workplace will also become routine as employers attempt to create a drug- and alcohol-free working environment.

## Testing your knowledge

| | |
|---|---|
| heart attack | blockage or bursting of a blood vessel in the brain that may cause brain damage |
| digital sphygmomanometer | chest pain that indicates that the heart muscle is not getting enough oxygen |
| stroke | high tech instrument that is used to measure blood pressure |
| angina | damage to heart muscle that results from blockage of a vessel and lack of oxygen |

**Figure 2.24**

**1** Copy and complete figure 2.24 using arrows to connect the words with their meanings. (4)

**2** a) What effect does exercise have on blood pressure? (1)
   b) Give THREE unhealthy aspects of lifestyle that cause high blood pressure. (3)
   c)  Name ONE problem that can be caused by low blood pressure. (1)

**3** Copy and complete table 2.9 using the following answers:
   anaemia, high concentration, high number, iron, presence, sugar. (6)

**4** a) A and B are two different blood groups. Name the other two. (1)
   b) Why would it be a serious mistake to give a person with blood group A a transfusion of blood group B? (1)

| aspect of blood tested | abnormal result indicating a problem | medical condition indicated |
|---|---|---|
| antibodies | | viral infection |
| | low content | |
| | | diabetes |
| white blood cell count | | leukaemia |

**Table 2.9**

## Applying your knowledge

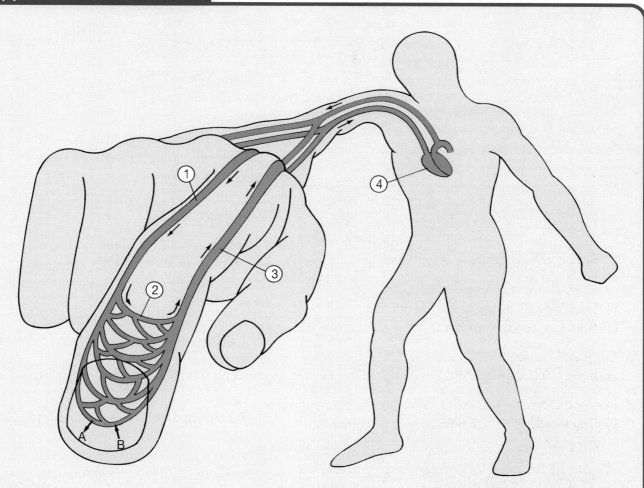

**Figure 2.25**

**1** Figure 2.25 shows part of the human circulatory system.

a) (i) Name structures 1–4.

  (ii) Which of these contain valves? (6)

b) (i) Name TWO substances that could be passing from the blood along arrow A to the living cells below the fingernail.

  (ii) Name ONE substance that could be passing along arrow B from the living cells to the blood. (3)

**2** Table 2.10 shows a set of results for three pupils taking several readings of their resting pulse rates.

a) (i) Calculate the average resting pulse rate as beats per 15 seconds for each pupil.

| pupil | resting pulse rate (beats/15 seconds) | | | | |
|-------|---------|---------|---------|---------|---------|
|       | trial 1 | trial 2 | trial 3 | trial 4 | trial 5 |
| Sandra | 18 | 17 | 16 | 16 | 18 |
| Rubina | 19 | 22 | 21 | 20 | 23 |
| Nicola | 16 | 14 | 20 | 15 | 15 |

**Table 2.10**

(ii) Calculate the average resting pulse rate as beats per minute for each pupil. (6)

b) (i) Who has the lowest resting pulse rate?
   (ii) What further information would you need before you could say that she is the fittest of the three? (2)

3 The bar graph in figure 2.26 shows the rate of blood flow in various parts of a student's body under differing conditions of exercise.

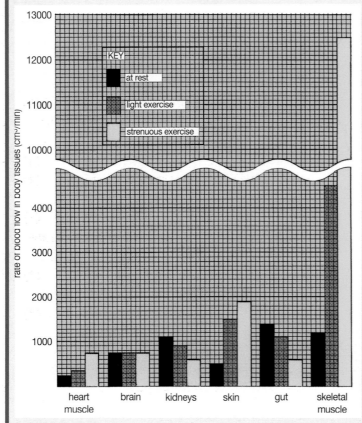

**Figure 2.26**

a) Using this information, copy and complete table 2.11. (5)

b) What effect does increasingly strenuous exercise have on the rate of blood flow in the:
   (i) brain;
   (ii) skeletal muscle;
   (iii) gut? (3)

c) Which other part(s) of the body shows the same trend in response to increase in exercise as the:
   (i) gut;
   (ii) heart muscle? (2)

4 The steps in the procedure used to measure length of recovery time after vigorous exercise are given below but in the wrong order.
   (1) Exercise vigorously for three minutes.
   (2) Calculate the time required for pulse rate to return to normal.
   (3) Measure normal pulse rate before starting to exercise.
   (4) Take pulse rate at one-minute intervals until normal pulse is recorded.

a) Arrange steps (1)–(4) in the correct order.

b) One afternoon this procedure was used to compare the recovery times of several college students. Some students had been playing sports at lunchtime and had missed lunch. Some students had been studying until late the night before while others had gone to bed early.

| part of body | rate of blood flow (cm³/minute) | | |
|---|---|---|---|
| | at rest | | strenuous exercise |
| heart muscle | | 350 | 750 |
| | 750 | 750 | |
| kidneys | | 900 | 600 |
| | 500 | | 1900 |
| gut | 1400 | 1100 | |
| skeletal muscle | 1200 | | |

**Table 2.11**

| time from start (min) | condition | pulse rate (beats/min) | |
|---|---|---|---|
| | | **Jane** | **Karen** |
| 0 | body at rest | 76 | 68 |
| 5 | exercise started now | 76 | 68 |
| 10 | exercise continuing | 124 | 106 |
| 15 | exercise stopped now | 124 | 106 |
| 20 | body at rest | 108 | 68 |
| 25 | body at rest | 92 | 68 |
| 30 | body at rest | 76 | 68 |
| 35 | body at rest | 76 | 68 |

**Table 2.12**

(i) Explain why the investigation is not a fair test.

(ii) Suggest how it could be adapted to make it fair. (2)

**5** Table 2.12 shows a set of results from an experiment where two girls had their pulse rates taken before, during and after a period of exercise.

a) Plot the data as two line graphs on the same sheet of graph paper. Put time on the horizontal (x) axis. (4)

b) What was Jane's resting pulse rate? (1)

c) What was Jane's pulse rate after 10 minutes of exercise? (1)

d) How long did it take for Jane's pulse rate to return to normal after exercise had stopped? (1)

e) What was Karen's resting pulse rate? (1)

f) What was Karen's pulse rate after 10 minutes of exercise? (1)

g) How long did it take for Karen's pulse rate to return to normal after exercise had stopped? (1)

h) Who recovered more quickly after exercise? (1)

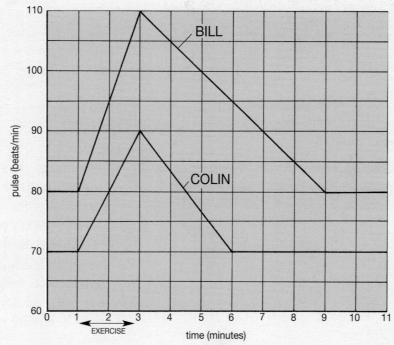

**Figure 2.27**

**6** The graph in figure 2.27 charts the pulse rates of two boys doing the same exercise for 120 seconds. The fitness of the heart can be measured by following pulse rate before, during and after exercise and then using the information to calculate **fitness index** using the formula:

$$\text{fitness index} = \frac{\text{time of exercise (seconds)}}{\text{increase in pulse} \times \text{recovery time (minutes)}}$$

a) (i)  For how long did Colin exercise (in seconds)?

(ii) By how many beats/minute did his pulse increase?

(iii) How long was his recovery time (in minutes)?

(iv) Calculate Colin's fitness index using the above formula. (4)

b) Repeat the steps given in a) for Bill. (4)

c) Refer to table 2.13 and describe each boy's level of fitness. (2)

| fitness index | level of fitness |
|---|---|
| 3 or more | excellent |
| 2–2.9 | very good |
| 1–1.9 | good |
| 0.5–0.9 | fair |
| 0.4 or less | unfit |

**Table 2.13**

**7** Table 2.14 shows how blood pressure changed as the blood went round the body of an adult man.

| blood pressure (mm Hg) | location of blood |
|---|---|
| 105 | leaving the heart |
| 90 | leaving the arteries |
| 15 | leaving the capillaries |
| 1 | leaving the veins |

**Table 2.14**

a) Present the information as a bar chart. (3)

b) (i)  What drop in pressure occurred between the time when the blood left the man's heart and the time when it left the arteries?

(ii) What drop in pressure occurred between the time when the blood left the arteries and the time when it left the capillaries?

(iii) By how many times was the pressure drop you gave as your answer to (ii) greater than that for (i)? (3)

c) Blood in veins is at a very low pressure. What stops the blood in leg veins flowing back down to the feet when the person is standing up? (1)

**8** In the 1990s, a survey was carried out in several countries to find out how many people per 100 000 of the population died from heart disease. In Japan, 10 women and 52 men died of heart disease, whereas in Germany the numbers were higher at 64 women and 246 men per 100 000. Scotland had 140 women and 510 men dying of heart disease while France had 35 female and 115 male deaths per 100 000 population.

a) Present the information in the above passage as a table. (3)

b) How many more men than women died of heart disease per 100 000 population in France? (1)

c) How many more Scottish men than French men died of heart disease per 100 000 population? (1)

d) (i)  What was the total number of deaths per 100 000 population for Japan?

(ii) What was the total number of deaths per 100 000 population for Germany?

(iii) By how many times was Germany's total number of deaths per 100 000 population greater than that of Japan? (3)

e) What is the ratio of Scottish women to Japanese women that died of heart disease? (1)

| age (years) | sex | upper blood pressure at rest (mm Hg) |
|---|---|---|
| 5 | female | 92 |
|   | male | 92 |
| 20 | female | 116 |
|   | male | 123 |
| 35 | female | 124 |
|   | male | 127 |
| 55 | female | 138 |
|   | male | 138 |

**Table 2.15**

**9** Table 2.15 shows average upper blood pressure readings for people of four different ages.

a) Draw a bar graph of this information using different colours for male and female. (4)

b) What happens to blood pressure as people get older? (1)

c) (i) At which ages did the blood pressure of males differ from those of females?
(ii) Was it higher or lower in men? (2)

**10** The following questions refer to table 2.6 on page 30.

a) To which blood groups can people with blood group A donate blood? (1)

b) From which blood groups can people with blood group B receive blood? (1)

c) Which blood group can donate blood to all the other blood groups? (1)

d) Which blood group can receive blood from all the other blood groups? (1)

**11** Table 2.16 shows the percentage distribution of blood groups in the UK.

a) Present the Scottish figures as a bar graph. (3)

b) Copy and complete the table. (6)

| blood group type | percentage | | | | |
|---|---|---|---|---|---|
| | UK region | | | | UK average |
| | Scotland | England | N. Ireland | Wales | |
| A | 34 | 42 | 26 | 38 | 35 |
| B | 11 | 8 | | 10 | |
| AB | 3 | 3 | 7 | | |
| O | 52 | | 32 | 49 | |

**Table 2.16**

## HYPERTENSION CAN KILL

A person's blood pressure can be taken using a stethoscope and a mercury manometer or a digital sphygmomanometer. If the lower blood pressure reading is found to be more than 90 mm Hg, this indicates high blood pressure. When a person's blood pressure is high all the time, they are said to be suffering from **hypertension**.

This condition is very bad for the health because it can lead to one or more of the following problems:

- Small blood vessels may burst under pressure. If these are blood capillaries in the brain, this causes a stroke.

- The walls of the arteries gradually become thicker under pressure. This makes it more difficult for blood to flow through them and so blood pressure rises still further.
- The inner lining of the arteries becomes damaged under pressure. When this happens, fatty deposits build up and leave less room for blood to flow (see figure 2.28).

As the years go by and fatty deposits build up, less and less space is left for blood to flow through. The narrow space that remains can easily be blocked by a blood clot. If the artery that takes blood to the heart wall becomes blocked in this way, the person suffers a **heart attack**.

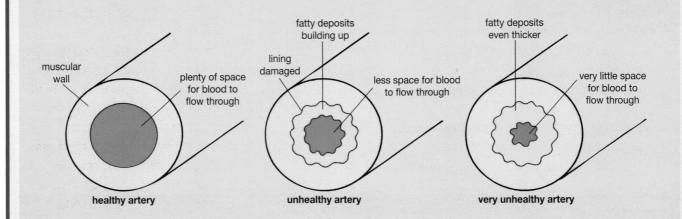

**Figure 2.28**

**12** Read the passage and answer the following questions.

a)  Identify the high-tech and the low-tech instruments named in the passage. (2)

b) If a person's lower blood pressure reading is always at least 95 mm Hg, what condition are they said to be suffering from? (1)

c) Why is someone with permanent high blood pressure at risk of suffering a stroke? (1)

d) Why does gradual thickening of artery walls raise blood pressure still further? (1)

e) When the inner lining of the artery leading to the heart becomes damaged under pressure, this sets off a chain of events that can lead to a heart attack. Represent this chain of events as a flow diagram. (4)

f) Suggest why insurance companies often insist on a person having a medical check-up before agreeing to give the person life insurance cover. (2)

## What you should know

| | | |
|---|---|---|
| age | fat | pressure |
| alcohol | fitness | pulse |
| anaemia | group | pulsometer |
| angina | heart | pumps |
| antibodies | high | recovery |
| arteries | leukaemia | sex |
| capillaries | link | short |
| carbon dioxide | low | stethoscope |
| digital | manometer | stopwatch |
| disease | muscular | strokes |
| excess | nutrients | sugar |
| exercising | overweight | veins |
| failure | oxygen | wastes |

**Table 2.17** *Word bank for chapter 2*

1 The circulatory system is made up of the _____ and the blood vessels. The heart is a _____ organ that _____ blood around the body.

2 The three main types of blood vessel are arteries, veins and _____. Blood is carried away from the heart in _____ and back to the heart in _____. Capillaries are tiny blood vessels found in body tissues. They form a _____ allowing blood to flow from arteries to veins.

3 Blood carries _____, oxygen, _____ and wastes round the body. Capillaries are thin-walled and therefore allow nutrients and _____ in the blood to pass out to the tissues and carbon dioxide and _____ to pass from the tissues into the blood.

4 Each time the heart beats, it pushes blood into the arteries making them swell. This movement is called _____. It can be measured using high-tech instruments such as a _____ or a heart rate monitor or low-tech instruments such as a stethoscope and a _____.

5 A person's pulse rate depends on several factors such as their size, _____, _____ and level of fitness. A high resting pulse rate may lead to heart _____.

6 The time taken for pulse rate to return to normal after exercise is called _____ time. A combination of _____ resting pulse rate and _____ recovery time may indicate high level of _____. Resting pulse rate and recovery time can be reduced by _____ regularly.

7 Blood that has been pumped into arteries is under pressure. This blood _____ can be measured using low-tech instruments called _____ and mercury _____ or a high-tech instrument called a _____ sphygmomanometer. A normal average upper value for blood pressure would be 120 and a lower value would be 80 (written as 120/80 for short). Readings over 160/90 indicate _____ blood pressure.

8 High blood pressure can be caused by: being _____, not taking enough exercise, eating a diet containing too much _____ or salt or drinking alcohol to _____. High blood pressure can lead to _____, heart attacks and _____. Low blood pressure can indicate heart _____.

9 Blood can be tested to detect several medical conditions. Infection is indicated by the presence of _____. _____ is indicated by a low iron content, diabetes by a high level of _____ and _____ by an abnormally high white blood cell count.

10 Before being used for a blood transfusion, blood has to be tested to find out its blood _____ since some blood groups clump in the presence of others instead of mixing freely. Blood is also tested for _____ concentration in cases of suspected drunk driving.

## 3 Healthy lungs

## Inhaled and exhaled air

Inhaled air means air that is breathed into the lungs. Exhaled air means air that is breathed out of the lungs.

### Activity

## Comparing the oxygen content of inhaled and exhaled air.

### You need
- length of tubing (rubber or plastic)
- disinfectant
- two gas jars
- trough of water
- beehive shelf
- deflagrating spoon with candle attached
- lid for gas jar
- stopclock

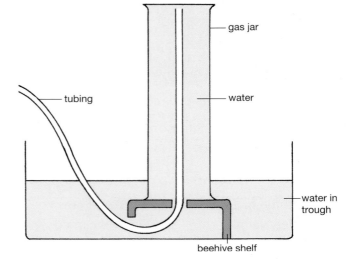

**Figure 3.1**  *Setting up a gas jar*

### What to do

1 Using disinfectant, clean the end of the tube that you are going to put in your mouth.
2 Set up the apparatus as shown in figure 3.1.
3 Blow into the tube and fill the gas jar with exhaled air.
4 Pull out the tube, fit the lid to the gas jar under water and then remove the jar of exhaled air from the trough.
5 Light the candle on the deflagrating spoon.
6 Slip the lid off the gas jar and quickly insert the spoon and burning candle (see figure 3.2).
7 Time how long the candle flame keeps burning.
8 Place the lid on the second gas jar which contains fresh air. (*Note*: Fresh air is the same as inhaled air.)
9 Repeat steps 5, 6 and 7 using inhaled air.
10 Answer the following questions:
   a) In which type of air (inhaled or exhaled) did the candle burn for the shorter time?
   b) In which type of air (inhaled or exhaled) did the candle burn for the longer time?

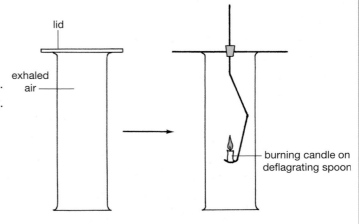

**Figure 3.2**  *Testing exhaled air for oxygen*

   c) (i) A flame is only able to burn when sufficient oxygen is present. Which type of air must have contained more oxygen? Which type of air must have contained less oxygen?
   (ii) From being breathed in to being breathed back out again, the air has lost some of its oxygen. Suggest where this oxygen has gone.

## Comparing the carbon dioxide content of inhaled and exhaled air

### You need

- two labels
- two boiling tubes
- stand to take boiling tubes
- lime water
- two rubber stoppers fitted with tubing (as shown in figure 3.3)
- disinfectant

### What to do

1 Label the boiling tubes **I** (for inhaled air) and **E** (for exhaled air).

2 Place the tubes (side by side) in the stand and one-third fill each with lime water. (*Note*: This volume must be equal in the two tubes so that the experiment is fair and a valid comparison of the two tubes can be made.)

3 Clean the end of each plastic tube mouthpiece with disinfectant.

4 Insert the rubber stoppers into the tubes.

5 Gently breathe in through tube **I**'s mouthpiece to a count of ten.

6 Quickly change over to tube **E**'s mouthpiece and breathe out to a count of ten.

7 Repeat steps 5 and 6 twice.

8 Answer the following questions:

　a) Which type of air (inhaled or exhaled) made the lime water turn very cloudy?

　b) Which type of air (inhaled or exhaled) made the lime water turn slightly cloudy?

　c) (i) When lime water turns cloudy, this shows that carbon dioxide ($CO_2$) is present. Which type of air must have contained more $CO_2$? Which type of air must have contained less $CO_2$?

　　(ii) From being breathed in to being breathed out again, the air has gained extra $CO_2$. Suggest where this $CO_2$ has come from.

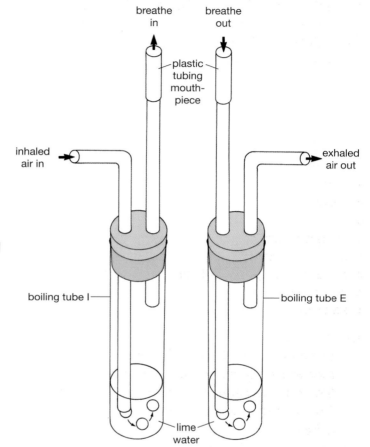

**Figure 3.3** *Testing air for carbon dioxide*

Activity

## Examining a sheep's lungs

### Information
This activity is intended to be carried out as a demonstration by the teacher.

### Teacher needs
- disposable gloves
- sheep's lungs and respiratory system
- dissecting board
- bicycle pump
- glass trough of water
- dissection instruments

### Pupils may need
- disposable gloves

### What to do
1 Put on disposable gloves if you intend to handle the sheep's lungs.
2 Gently prod a lung with your finger to feel its spongy texture.

3 Hold the windpipe in your hand and squeeze it gently to try to close it.
4 Watch as the teacher:
- points out where the base of the windpipe divides into two tubes each leading to a lung. Each of these tubes is called a bronchus (plural = bronchi).
- attaches a bicycle pump to the end of the windpipe and pumps air into the lungs to inflate them. (Also see figure 3.4.)
- tries to make one of the inflated lungs sink in the trough of water.
- cuts lengthways down through the windpipe's tough rings of supporting material and opens the windpipe to show its shiny inner surface.
- cuts off a lung and dissects it to show the bronchus dividing into smaller tubes called bronchioles. These end in microscopic air sacs surrounded by blood capillaries.

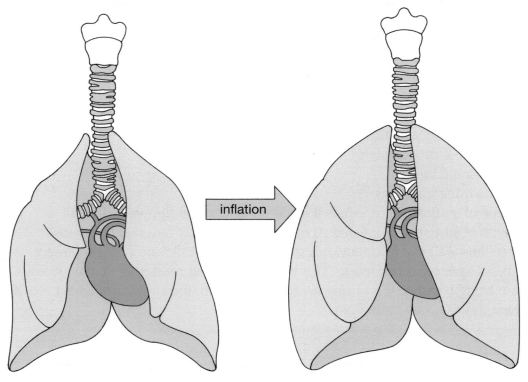

inflation

**Figure 3.4**  *Inflating lungs with air*

## Lungs and breathing

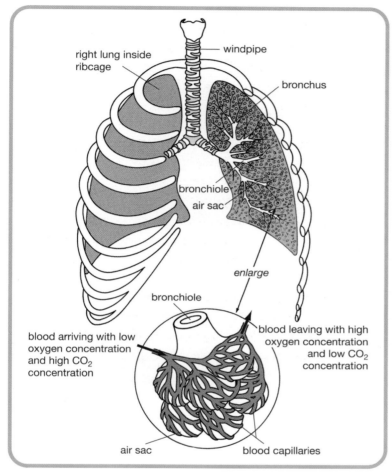

The **lungs** (see figure 3.5) are the organs of **gas exchange**. Air enters the breathing system by the nose or mouth. The air passes through the voice box, the **windpipe**, the **bronchi** and the **bronchioles** on its way to the millions of **air sacs** at the ends of the bronchioles.

### Function of the lungs

Each air sac is thin-walled and surrounded by tiny **blood vessels** (capillaries). It is here that gas exchange takes place. **Oxygen** in inhaled air passes from the air into the bloodstream. **Carbon dioxide** passes in the opposite direction from the bloodstream into the air sacs ready to be breathed out in exhaled air.

**Figure 3.5** *Lung structure*

### Activity

## Examining a torso model and radiographs (X-rays)

### Information
This activity is intended as a demonstration carried out by the teacher.

### Teacher needs
- human torso model
- radiographs (X-rays) of human chest

### What to do
1 Take your turn of handling and assembling the organs that make up the human chest in the model.
2 View radiographs (X-rays) of the human chest and figure 3.6. Try to identify: right and left lung, heart, windpipe, bronchi and rib cage.

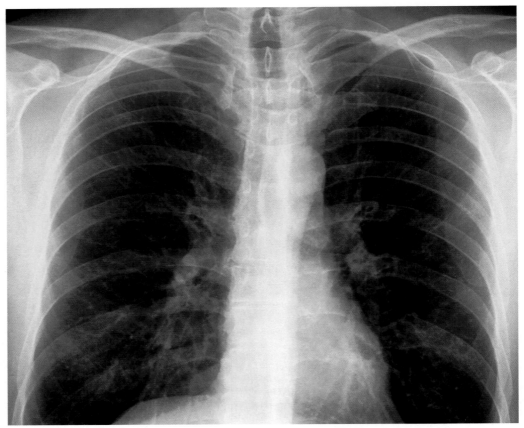

**Figure 3.6**  *An X-ray of healthy lungs*

## Breathing rate

**Breathing rate** is the number of breaths taken per minute. Each full breath has two parts: **inhalation** (breathing in) and **exhalation** (breathing out).

A person's breathing rate varies depending on their *level of activity*. The more active they are, the greater their breathing rate. This is because they need to breathe more often to take in the oxygen needed for energy release.

Breathing rate varies from person to person depending on several factors:

- **age** (it is greater in younger people who are more active than older people);
- **sex (gender)** (it is slightly lower, on average, in women);
- **level of fitness and state of health** (it is lower in fitter, healthier people).

## Activity

# Investigating the effect of exercise on breathing rate

*You need*
- stopwatch
- calculator

| time from start (min) | situation | breathing rate (breaths/min) |
|---|---|---|
| 0 | at rest | |
| 3 | immediately after completing exercise | |
| 4 | 1 minute after exercise | |
| 5 | 2 minutes after exercise | |
| 6 | 3 minutes after exercise | |
| 7 | 4 minutes after exercise | |
| 8 | 5 minutes after exercise | |
| 9 | 6 minutes after exercise | |
| 10 | 7 minutes after exercise | |

**Table 3.1**   Breathing rate results

*What to do*
**1** Prepare a copy of table 3.1.
**2** Using the stopwatch, time yourself sitting quietly for two minutes to allow your breathing rate to settle at a steady pace.
**3** Start the stopwatch and measure your breathing rate for one minute while continuing to sit quietly. (*Note*: Each breath involves breathing in *and* out.)
**4** Repeat step 3 twice and calculate your average breathing rate at rest. (*Note*: Taking several readings and calculating an average gives an overall result that is *more reliable*.) Enter the result in the table at 0 minutes from the start.
**5** Exercise vigorously by running on the spot for three minutes.
**6** Stop exercising and immediately measure your breathing rate. Record your result in the table at three minutes from the start.
**7** Continue measuring your breathing rate every minute for a further seven minutes and enter each result in the table.
**8** Draw a line graph of your results with time on the horizontal (x) axis.
**9** Answer the following questions:
   a) What effect did exercise have on breathing rate?
   b) How long did it take for your breathing rate to return to its resting value?
   c) How long was your recovery time?

## Exercise and breathing rate

**Exercise** makes breathing rate increase. The person takes *more* breaths per minute than they do at rest. They also take *deeper* breaths. Both of these changes to breathing help to ventilate the lungs thoroughly during and after exercise. This increases the rate of gas exchange between the air in the air sacs and the blood flowing through the lungs. As a result, the person is able to take in oxygen to replace that used up during exercise and to get rid of the carbon dioxide formed during exercise.

## Recovery time and fitness

The **rate** and **depth** of breathing return to normal during the **recovery time** after exercise. The *shorter* the recovery time, the *fitter* the person.

## Testing your knowledge

1 Copy and complete table 3.2 using the words *more* or *less*. (4)

2 Match boxes 1–6 in figure 3.7 with the words in the following word bank: air sac, bronchiole, bronchus, lung, space for heart, windpipe. (6)

| type of air | oxygen content | carbon dioxide content |
|---|---|---|
| inhaled | | |
| exhaled | | |

**Table 3.2**

3 Make a flow chart to show the route taken by a molecule of oxygen that enters the lungs in inhaled air and ends up in the bloodstream. (4)

4 Decide whether each of the following statements is true or false and use T or F to indicate your choice. Where a statement is false, give the word that should have been used in place of the word in **bold** print. (6)

a) A function of the lungs is to take in **oxygen** from the air and pass it into the blood.

b) A function of the lungs is to remove carbon dioxide from the **blood**.

c) Breathing rate is measured in **beats**/minute.

d) Exercise results in breathing becoming **slower** and deeper.

e) The time taken for the body to return to its resting state after exercise is called **recovery** time.

f) A **long** recovery time indicates that the person is fit.

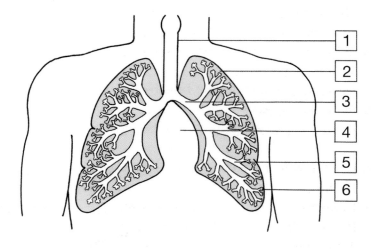

**Figure 3.7**

# Physiological measurements of the lungs

## Vital capacity

**Vital capacity** is a measure of the *maximum* volume of air that a person can breathe out after they have taken in as deep a breath as possible. Vital capacity varies from person to person. It depends on factors such as:

- **body size** (the larger the person's build, the greater their vital capacity);
- **age** (vital capacity reaches a maximum by age 20 approximately);
- **sex (gender)** (the average woman has a smaller vital capacity than the average man);
- **fitness** (vital capacity increases with increased level of fitness).

## Activity

### Measuring vital capacity (low-tech method)

*You need*
- beaker (1 litre, plastic)
- waterproof marker
- bell jar
- disinfectant
- length of tubing (plastic or rubber)
- tank of water

*What to do*
1 Using the beaker and marker, give the bell jar a scale.
2 Use disinfectant to clean the end of the tube that you are going to blow through.
3 Set up the equipment as shown in figure 3.8.
4 Take in as deep a breath as possible. (This is called **maximum inspiration**.)
5 Place the cleaned end of the tube in your mouth and breathe out as much air as you can.
6 Use the scale to measure your vital capacity in litres.
7 Express this value in: (i) millilitres (ml); (ii) cubic centimetres (cm$^3$).
   [*Note*: 1 litre (l) = 1000 millilitres (ml)
   1 millilitre (ml) = 1 cubic centimetre (cm$^3$)]

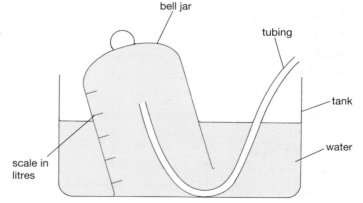

**Figure 3.8** *Measuring vital capacity*

## Activity

## Measuring vital capacity (high-tech method)

### Information

The **spirometer** (see figure 3.9) is a piece of high-tech equipment used to measure vital capacity in cm³ without using water.

### You need

- spirometer (spiropet)
- sterile mouthpiece

| attempt number | vital capacity (cm³) |
|---|---|
| 1 | |
| 2 | |
| 3 | |

**Table 3.3** *Vital capacity results*

**Figure 3.9** *Spiropet spirometer*

### What to do

1 Prepare a copy of table 3.3.
2 Fit a sterile mouthpiece on to the nozzle of the spirometer.
3 Turn the ring until the pointer is at zero on the scale.
4 Hold the spirometer by its base with the mouthpiece sticking out horizontally as shown in figure 3.10. (Take care not to cover the small holes with your hand.)
5 Breathe in deeply until your lungs are full.
6 Breathe out completely into the spirometer over a period of about five seconds.
7 Read the scale and record this attempt in your table.
8 Repeat steps 2–7 twice.
9 Dispose of the mouthpiece as instructed.

*Note*: Normally in an experiment when repeat readings have been taken, an average is calculated to give an overall result that is more reliable. However, vital capacity is an exception to this rule. The *highest value* is your vital capacity.

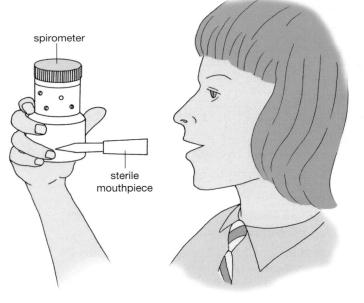

**Figure 3.10** *Using a spirometer*

## Tidal volume

**Tidal volume** is the volume of air breathed in *or* out of the lungs in one normal breath. Tidal volume varies from person to person depending on their **body size**, **age**, **sex** and **level of fitness**.

### Activity

## Measuring tidal volume

### You need
- breath volume kit
- disinfectant
- elastic band

### What to do
1 Clean the mouthpiece, valves, coupler and T-piece with disinfectant.
2 Assemble the parts as shown in figure 3.11. (Make sure that the valve tabs marked I face in the same direction and that the top of the bag is secured to the coupler with the elastic band.)
3 Seal your mouth round the mouthpiece.
4 Hold your nose and take three normal breaths. (*Note*: One breath = inhale + exhale).
5 Squeeze the air down into the end of the bag (see figure 3.12) and measure its volume.
6 Calculate your tidal volume by dividing the volume of air in the bag by three.

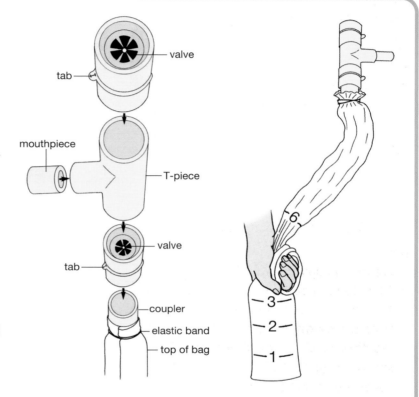

**Figure 3.11** *Assembling a breath volume kit*

**Figure 3.12** *Collecting a volume of exhaled air*

## Vital capacity and tidal volume

The volume of air entering and leaving the lungs can also be measured using the apparatus shown in figure 3.13. This is called a trace-drawing spirometer. It can be used to measure both tidal volume and vital capacity.

The floating drum is connected to a pen that draws a line on graph paper attached to a slowly revolving drum. When the person breathes in, the line made by the pen goes up. When the person breathes out, the line made by the pen goes down. The graph in figure 3.14 shows a detailed version of the line drawn on the drum when the person attached to the apparatus was breathing in and out.

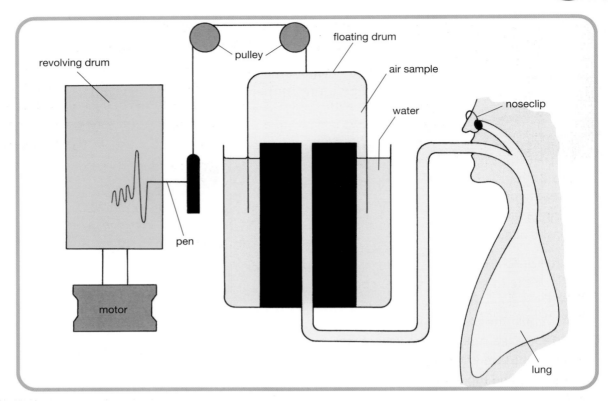

**Figure 3.13** *Trace-drawing spirometer*

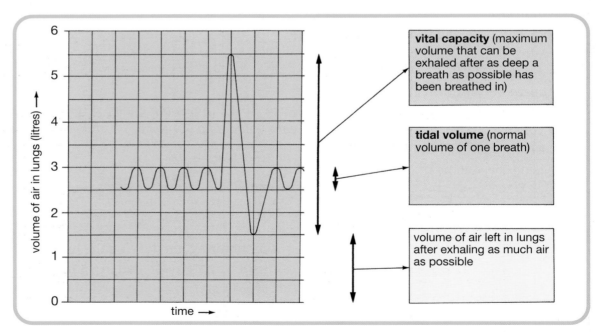

**Figure 3.14** *Trace made by spirometer*

## Peak flow

**Peak flow** is the maximum rate at which air can be forced out of the lungs. It varies from person to person depending on their **body size**, **age**, **sex** and **level of fitness**.

**Figure 3.15** *Peak flow meter*

Peak flow can be measured using a peak flow meter (see figure 3.15). This equipment is often used by medical staff to check the state of a patient's airways. Sufferers of asthma have airways that become narrower under certain circumstances. This means that at these times an asthmatic cannot force air out of their lungs as quickly as a non-sufferer. Use of a peak flow meter therefore helps to detect asthma. The sufferer can be given an inhaler to help to manage the condition (see also question 7 on page 61).

## Activity

## Measuring peak flow

*You need*
- peak flow meter
- sterile mouthpiece

*What to do*
1 Prepare a copy of table 3.4.
2 Follow the procedure shown in figure 3.16.
3 Find out your peak flow rate by taking the *highest* value of three attempts (*not* an average).

| attempt number | peak flow (litres/min) |
|---|---|
| 1 | |
| 2 | |
| 3 | |

**Table 3.4** *Peak flow results*

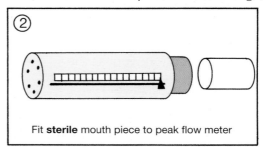
Push pointer back to zero on scale

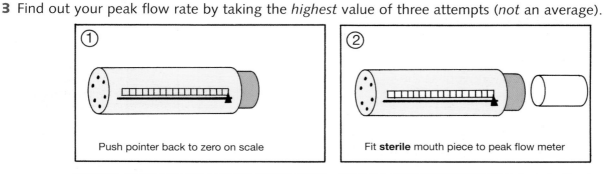
Fit **sterile** mouth piece to peak flow meter

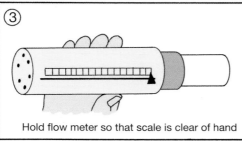
Hold flow meter so that scale is clear of hand

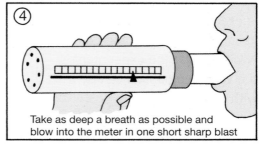
Take as deep a breath as possible and blow into the meter in one short sharp blast

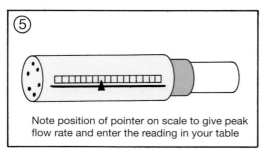
Note position of pointer on scale to give peak flow rate and enter the reading in your table

Repeat steps 1–5 twice

**Figure 3.16** *Measuring peak flow*

## Effects of smoking on health

Activity

### Investigating the tar content of cigarettes

#### Information
This activity is intended as a demonstration by the teacher.

#### Teacher needs
- three U-tubes
- cotton wool
- rubber tubing
- beaker of crushed ice
- suction pump
- cigarettes (three brands)
- matches

| type of cigarette | tar content |
|---|---|
|  |  |
|  |  |
|  |  |

**Table 3.5**  *Cigarette tar results*

#### What to do
1 Prepare a copy of table 3.5.
2 Watch as:
  - the teacher sets up the smoking machine (see figure 3.17);
  - tar gathers in the cotton wool as the machine smokes the cigarette of the first brand;
  - a fresh U-tube is fitted and the procedure is repeated with a cigarette of the second brand. (*Note*: This U-tube must be exactly the same as the first one and contain the same mass of cotton wool so that the experiment is fair and a valid comparison can be made with the first U-tube.);
  - the procedure is repeated with a cigarette of the third brand.

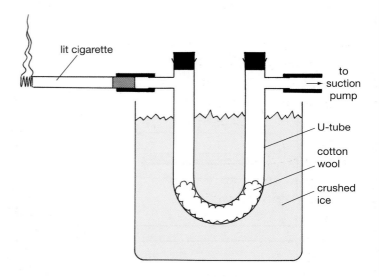

**Figure 3.17**  *Smoking machine*

3 Compare the three U-tubes and complete your table using the terms *light brown*, *medium brown* and *dark brown* to describe the tar content that gathered.
4 Answer the following questions:
  a) What was the one variable factor that was investigated in this experiment?
  b) Draw a conclusion from the results.
  c) Describe how you could use the smoking machine to investigate whether tips on cigarettes reduce the amount of tar breathed in by the smoker.

## Activity

# Demonstrating carbon monoxide in cigarette smoke

### Information

This activity is intended as a demonstration by the teacher.

### Teacher needs

- carbon monoxide detector (e.g. carbon monoxide alarm SF Detection Model 350 – see figure 3.18)
- smoking machine with pump and one-way valve (e.g. 'Smokey Sue')
- cigarettes
- matches

**Figure 3.18**  *Carbon monoxide detector*

### What to do

1 Watch as the teacher:
   - checks the carbon monoxide detector;
   - sets up the smoking machine with a lit cigarette (also see figure 3.19);
   - uses the hand pump to make the machine take several draws of the cigarette and allows the smoke to collect in the plastic tube sealed at the end with a small rubber stopper;
   - removes the rubber stopper and directs the smoke into the gas inlets on the carbon monoxide detector (see figure 3.20).

2 a) Describe what happened a few seconds after the cigarette smoke had entered the carbon monoxide detector.
   b) Draw a conclusion from this experiment.

**Figure 3.19**  *'Smokey Sue'*

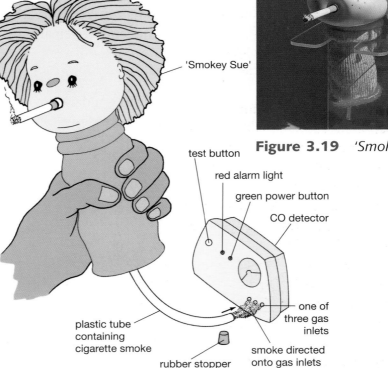

'Smokey Sue'

test button

red alarm light

green power button

CO detector

one of three gas inlets

smoke directed onto gas inlets

rubber stopper

plastic tube containing cigarette smoke

**Figure 3.20**  *Testing cigarette smoke for carbon monoxide (CO)*

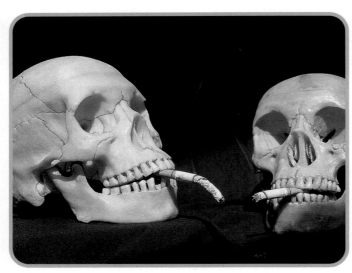

**Figure 3.21**  *Not long to go!*

## Smoking and health risks

Smoking seriously harms the health (see figure 3.21). Cigarette smoke contains **carbon monoxide**, **nicotine** and **tar**.

## Carbon monoxide

**Carbon monoxide** is a gas that has no taste or smell. However it is poisonous. Once it gets into the bloodstream, it combines with the red chemical (haemoglobin) present in red blood cells. This chemical is no longer able to pick up oxygen properly. So the person's heart has to work harder to supply the tissues with oxygen. Eventually this can lead to **heart disease**.

## Nicotine

**Nicotine** is a drug made by tobacco plants to protect themselves against insects. Once nicotine has entered the smoker's body, it stimulates the person's nervous system and needlessly increases their **pulse rate** and **blood pressure**. Nicotine is an **addictive** drug which makes the smoker crave cigarettes.

## Tar

**Tar** from tobacco is a dark brown, sticky substance that gathers inside the airways of the smoker's lungs. The tiny **hairs** (see close-up in figure 3.22) that normally sweep the airways clean are damaged by tar and nicotine. The tar contains hundreds of substances that have been shown to be harmful. Some of these are **carcinogens**. A carcinogen is a substance that can cause **cancer**.

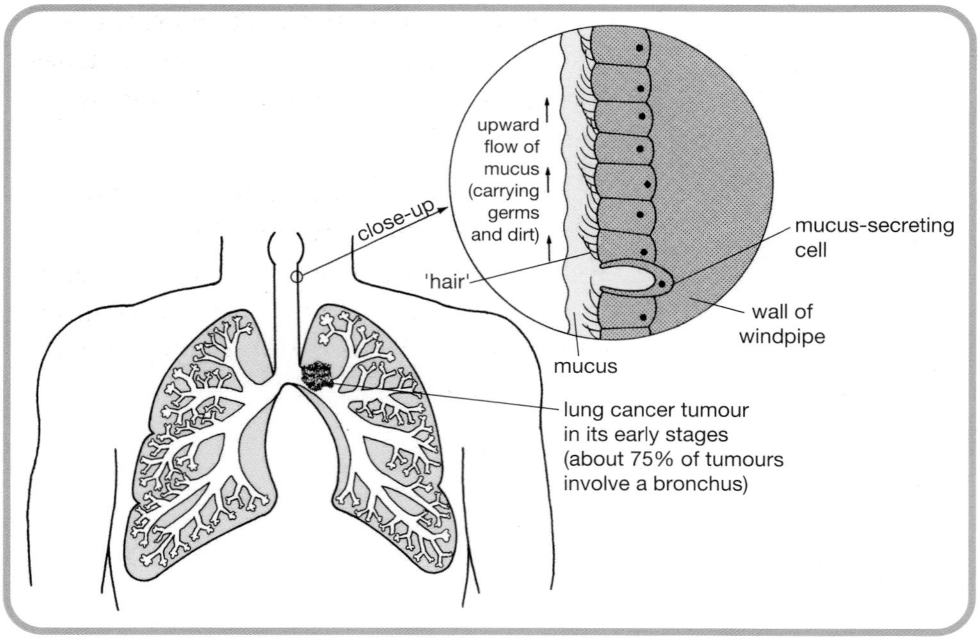

upward flow of mucus (carrying germs and dirt)

close-up

'hair'

mucus-secreting cell

wall of windpipe

mucus

lung cancer tumour in its early stages (about 75% of tumours involve a bronchus)

**Figure 3.22**  *Lungs affected by smoking*

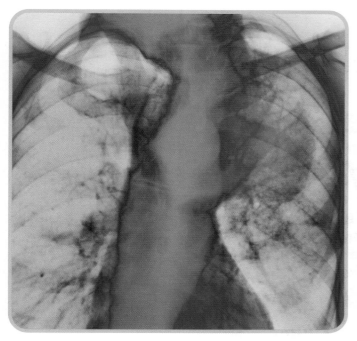

**Figure 3.23** *X-ray of a cancerous lung*

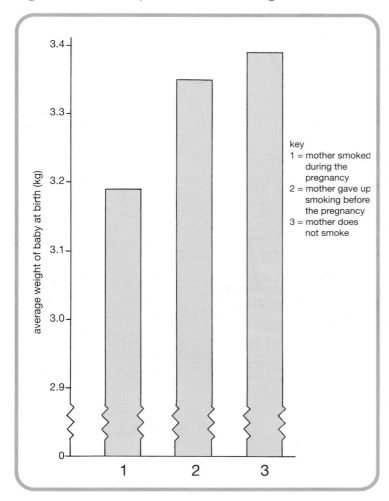

**Figure 3.24** *Effects of smoking on an unborn baby*

## Cancer

**Cancer** is an uncontrolled growth of cells. The most common form of lung cancer begins in the wall of an airway as a small lump of cells (a **tumour** – see figure 3.22) which increases in size over a period of many years, often without being discovered. To be found in time for successful treatment, lung tumours need to be picked up in their early stages by examining X-rays of the person's chest. Nine out of ten people who die of lung cancer (see figure 3.23) are smokers. Cigars and pipe tobacco are less harmful to the lungs because less smoke is breathed in. However they lead to cancers of the lips, mouth and voicebox.

## Smoking and pregnancy

Smoking during pregnancy slows down the growth of the unborn child. This happens for the following reasons.

- The baby gets *less oxygen* than it needs because the mother's blood is polluted with carbon monoxide and is only carrying about 90% of the oxygen that it should be taking to the baby.
- Nicotine in the mother's bloodstream *reduces* the amount of *glucose* that reaches the growing baby's tissues including its brain cells.

On average, babies born to mothers who smoke are *lighter* (see figure 3.24) and less healthy than those of non-smokers. Evidence also suggests that, on average, babies of non-smokers grow up to be more intelligent than those of smokers.

Cigarette packets carry a variety of messages to remind smokers of the risks that they are taking (see figure 3.25).

**Figure 3.25** *Risk reminders on cigarette packets*

## Testing your knowledge

1 Copy figure 3.26 and complete it by linking the measurements to their meanings by arrows. (3)

| tidal volume | | maximum rate at which air can be forced out of the lungs |
| peak flow | | maximum volume of air that can be breathed out in one breath after a maximum inspiration |
| vital capacity | | volume of air breathed in or out of the lungs in one normal breath |

**Figure 3.26**

2 a) Which of the following is a respiratory condition which always involves wheezing and difficulty in breathing?

  **A** anaemia  **B** anorexia  **C** asthma  **D** allergy (1)

  b) Which of the measurements in question 1 gives information that helps to detect this condition? (1)

  c) Give ONE way in which a sufferer could get fast relief if they felt an attack coming on. (1)

3 a) Which gas present in cigarette smoke reduces the ability of the blood to carry oxygen round the body? (1)

  b) Smoking increases the risk of several diseases. Name TWO of these. (2)

  c) What effect does smoking when pregnant usually have on the baby's birth weight? (1)

## Applying your knowledge

1 Figure 3.27 shows a set of equipment used to compare the carbon dioxide content of inhaled (I) and exhaled (E) air. However two changes need to be made to tube E to allow a fair comparison.
Make a labelled diagram of tube E correctly set up. (3)

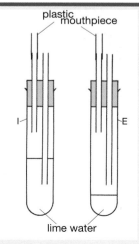

**Figure 3.27**

| state of body | volume of air in each breath (cm³) | number of breaths taken each minute | total volume of air breathed in one minute (cm³) |
|---|---|---|---|
| at rest | 500 | 15 | X |
| after exercise | 1000 | 30 | Y |

**Table 3.6**

2 Table 3.6 shows data from a 15 year-old boy before and after exercise.
   a) Calculate the answers that should have been entered in boxes X and Y of the table. (2)
   b) What effect did exercise have on the boy's
      (i)   rate of breathing?
      (ii)  depth of breathing? (2)
   c) Following exercise, 4% of the air breathed in by the boy consisted of oxygen that successfully *passed into his bloodstream* and was not breathed back out again. Calculate the volume of oxygen entering his bloodstream per minute after exercise. (2)

3 Three 14-year-old pupils ran non-stop round a running track for 12 minutes to test their fitness.
   a) Joe ran at an average speed of 200 metres/minute (m/min).
      (i)   Calculate the total distance (in metres) that he covered in 12 minutes.
      (ii)  Refer to table 3.7 and state Joe's level of fitness. (2)

| total distance covered (m) | level of fitness |
|---|---|
| 1600 or less | very poor |
| 1601–2000 | poor |
| 2001–2400 | fair |
| 2401–2800 | good |
| 2801 or more | very good |

**Table 3.7**

   b) Nicole ran at an average speed of 220 m/min.
      (i)   Calculate the total distance (in metres) that she covered in 12 minutes.
      (ii)  Was Nicole more or less fit than Joe?
      (iii) Explain your answer. (3)
   c) Abdul covered a total distance of 2.7 kilometres in 12 minutes.
      (i)   State this total distance in metres and then state his level of fitness.
      (ii)  Calculate his average speed in m/min. (3)
   d) The maximum oxygen intake (MOI) which a person can manage is also a good indication of fitness. The higher the MOI, the fitter the person. Draw TWO conclusions from table 3.8. (2)

| age (years) | MOI of athlete (l/min) | MOI of non-athlete (l/min) |
|---|---|---|
| 20–29 | 4.8 | 3.2 |
| 30–39 | 4.6 | 2.9 |
| 40–49 | 4.0 | 2.7 |

**Table 3.8**

4 The boy shown in figure 3.28 is about to use the apparatus to measure his vital capacity.
   a) What is meant by the term *vital capacity*? (2)
   b) What should the boy do to the end of the tube before he puts it in his mouth? (1)
   c) State TWO changes that the boy will have to make to the equipment before he begins blowing. (2)

d) What will happen to the level of the water
   (i)   inside
   (ii)  outside the bell jar once he starts to breathe out? (2)

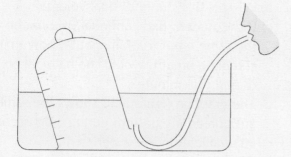

**Figure 3.28**

5 Table 3.9 gives average values for vital capacity.
   a) On average which sex has the higher vital capacity at any age? (1)
   b) At approximately what age does vital capacity reach its maximum in:
      (i)   males?
      (ii)  females? (2)
   c) The average 30 year-old man has a vital capacity equal to that of a male teenager of what age? (1)
   d) The average 15 year-old female has a vital capacity equal to that of an adult female of what age? (1)
   e) An average vital capacity of 4150 cm³ would be typical of an adult male of what age? (1)
   f) On average a woman's vital capacity is found to have decreased to below 2 litres by the time she is what age? (1)
   g) By how many cm³ is the average vital capacity of a 14 year-old boy greater than that of a 14 year-old girl? (1)
   h) By how many cm³ is the average vital capacity of a 15 year-old boy lower than that of a 33 year-old man? (1)

| age (years) | vital capacity (cm³) | |
| --- | --- | --- |
| | female | male |
| 4 | 600 | 700 |
| 5 | 800 | 850 |
| 6 | 980 | 1070 |
| 7 | 1150 | 1300 |
| 8 | 1350 | 1500 |
| 9 | 1550 | 1700 |
| 10 | 1740 | 1950 |
| 11 | 1950 | 2200 |
| 12 | 2150 | 2540 |
| 13 | 2350 | 2900 |
| 14 | 2480 | 3250 |
| 15 | 2700 | 3600 |
| 16 | 2700 | 3900 |
| 17 | 2750 | 4100 |
| 18 | 2800 | 4200 |
| 19 | 2800 | 4300 |
| 20 | 2800 | 4320 |
| 21 | 2800 | 4320 |
| 22 | 2800 | 4300 |
| 23 | 2790 | 4280 |
| 24 | 2780 | 4250 |
| 25 | 2770 | 4220 |
| 26 | 2760 | 4200 |
| 27 | 2740 | 4180 |
| 28 | 2720 | 4150 |
| 29 | 2710 | 4120 |
| 30 | 2700 | 4100 |
| 31–35 | 2640 | 3990 |
| 36–40 | 2520 | 3800 |
| 41–45 | 2390 | 3600 |
| 46–50 | 2250 | 3410 |
| 51–55 | 2160 | 3240 |
| 56–60 | 2060 | 3100 |
| 61 and over | 1960 | 2970 |

**Table 3.9**

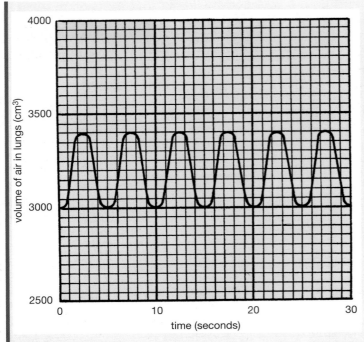

**Figure 3.29**

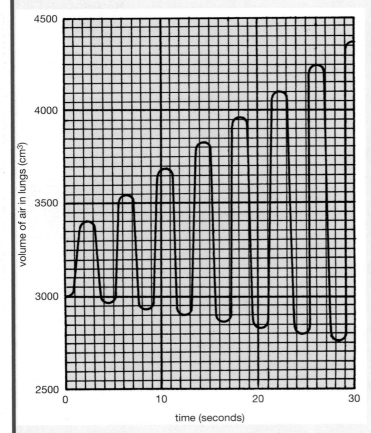

**Figure 3.30**

**6** The graph shown in figure 3.29 shows the result of a student breathing normally at rest while connected to a trace-drawing spirometer.

a) What is meant by the term *tidal volume*? (2)

b) What is this student's tidal volume? (1)

c) If the student had continued to breathe at the same rate for a full minute, how many complete breaths would she have taken during that minute? (1)

d) The graph in figure 3.30 shows the same girl's breathing pattern during a 30-second interval later in the experiment.

(i)   In what way have her rate and depth of breathing changed?

(ii)  Suggest what the girl was doing to make her breathing pattern change in this way. (3)

**7** Read the passage on page 61 and answer the following questions.

a) Give TWO common signs of untreated asthma. (2)

b) What is each of the following used for?
(i)   peak flow meter,
(ii)  inhaler. (4)

c) Draw a pie chart to represent the proportion of 9 year-old boys compared to that of 9 year-old girls who suffer asthma in the UK. (2)

d) Name TWO factors that can bring on an asthmatic attack without any unwanted particles being breathed in. (2)

e) (i)   Why are severe asthmatics often advised not to keep household pets?

(ii)  Give TWO other parts of a management plan that an asthmatic should follow. (3)

## MANAGING ASTHMATIC ATTACKS

Asthma is a respiratory condition in which the sufferer's air passages become narrower (see figure 3.31). This makes it more difficult for the person to breathe. In addition to suffering shortness of breath, the person tends to wheeze, especially after exercise. In extreme situations their chest becomes so tight that they feel as if they are going to suffocate.

A peak flow meter is used by a doctor to find out how well a patient's airways are working. It measures how fast air can be expelled from the lungs.

An asthmatic attack can be caused by an allergic reaction to dust mites, pollen, animal fur and certain foods such as peanuts and wheat. It can also be brought on by nervous tension. In the UK, boys below the age of 10 are twice as likely to be affected as girls of the same age. Asthma sufferers are able to control the condition if they follow a management plan. This includes:

- controlling house mites;
- avoiding foods to which they are allergic;
- warming up gently before exercise;
- using an inhaler (see figure 3.32) to take a drug that gives relief by making muscles in the airways relax and open the tubes wider.

Following this plan allows the person to play sports and lead a normal full life.

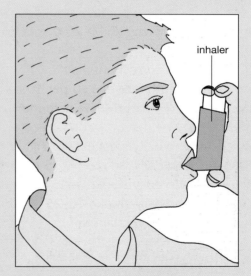

Figure 3.32

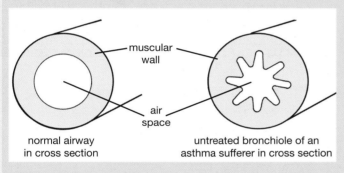

Figure 3.31

8 The chart in figure 3.33 on page 62 shows the peak flow readings recorded by a sufferer of asthma over one week. Each black box entry on the chart is the highest of three blows.
   a) How many times did the person use their peak flow meter
      (i) per day,
      (ii) per week? (2)
   b) What was the highest peak flow rate recorded? (1)
   c) What was the lowest peak flow rate recorded? (1)
   d) Copy and complete table 3.10 to show the days and times of the day that the person probably had to use their inhaler to relieve an asthmatic attack. (6)

| day | time of day | peak flow rate (l/min) before using bronchodilator |
| --- | --- | --- |
|  |  |  |
|  |  |  |

Table 3.10

**9** The bar chart shown in figure 3.34 is based on a survey of the number of cigarettes smoked daily and the death rates from coronary heart disease in British men.

a) An annual death rate of 100 per 100 000 from coronary heart disease is found for men aged below 45 who daily smoke
**A** 0 cigs **B** 1–15 cigs **C** 16–25 cigs **D** 26 or more cigs
(Choose ONE correct answer) (1)

b) An annual death rate of 360 per 100 000 from coronary heart disease is found for men aged 45–54 who daily smoke
**A** 0 cigs **B** 1–15 cigs **C** 16–25 cigs **D** 26 or more cigs
(Choose ONE correct answer) (1)

c) An annual death rate of 740 per 100 000 from coronary heart disease is found for men aged
**A** 45–54 who daily smoke 1–15 cigarettes
**B** 45–54 who daily smoke 16–25 cigarettes
**C** 55–64 who daily smoke 1–15 cigarettes
**D** 55–64 who daily smoke 16–25 cigarettes.
(Choose ONE correct answer.) (1)

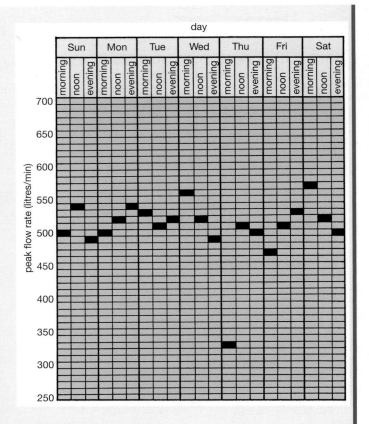

**Figure 3.33**

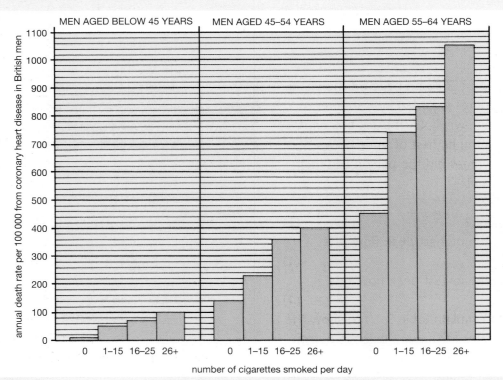

**Figure 3.34**

d) What is the annual death rate per 100 000 from coronary heart disease for 30 year-old men who smoke 15 cigarettes per day? (1)

e) What is the annual death rate per 100 000 from coronary heart disease for 50 year-old men who smoke 20 cigarettes per day? (1)

| group | deaths per 100 000 people from lung cancer |
|---|---|
| non-smokers | 11 |
| cigar smokers | 22 |
| pipe smokers | 33 |
| cigarette smokers | 110 |

**Table 3.11**

f) What is the annual death rate per 100 000 from coronary heart disease for 60 year-old men who smoke 30 cigarettes per day? (1)

**10** Table 3.11 shows the results of a survey into deaths from lung cancer by different groups of people.

a) Draw a bar chart of this information. (3)

b) By how many times is the number of deaths per 100 000 people greater for cigarette smokers than for non-smokers? (1)

c) (i) By how many times is the number of deaths per 100 000 people greater for pipe smokers than for non-smokers?

(ii) Suggest why fewer pipe smokers than cigarette smokers die of lung cancer. (2)

d) The figures in the table only refer to lung cancer.

(i) Predict the effect on the graph's cigar and pipe smoking bars if mouth and throat cancer had been included.

(ii) Explain your answer. (2)

**11** The six experiments in figure 3.35 were set up to investigate factors affecting the tar content of cigarettes.

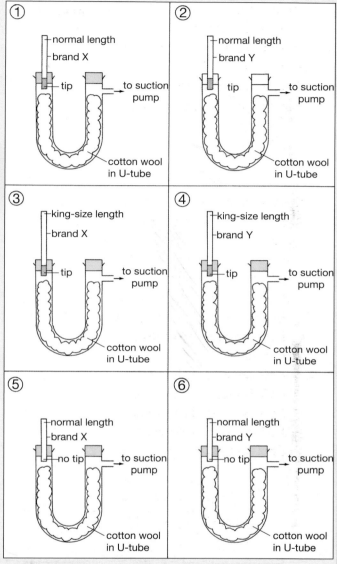

**Figure 3.35**

a) Which TWO experiments should be compared to find out the effect on tar content of *brand type* in normal-sized, tipped cigarettes? (1)

b) Which TWO experiments should be compared to find out the effect on tar content of *cigarette size* of tipped cigarettes of brand Y? (1)

c) Which TWO experiments should be compared to find out the effect on tar content of *presence or absence of tips* in normal-sized cigarettes of brand X? (1)

d) What factor could be investigated by comparing experiments 1 and 3? (1)

e) Why is a comparison of experiments 4 and 6 invalid? (1)

**12** The graphs in figure 3.36 show the percentage number of babies of different birth weights born in a hospital over a period of one year. Some of the mothers were smokers and some were non-smokers.

a) Out of 100 babies, how many of weight 3.1– 4.0 kg were born to mothers who did not smoke? (1)

b) Out of 100 babies, how many of weight 2.1– 3.0 kg were born to mothers who smoked? (1)

c) What percentage of babies born to non-smoker mothers weighed
   (i)   3.0 kg or less?
   (ii)  3.1 kg or more? (2)

d) What percentage of babies born to mothers who smoked weighed
   (i)   3.0 kg or less?
   (ii)  3.1 kg or more? (2)

e) (i)   Which type of mothers, on average, produced heavier babies?
   (ii)  Explain why. (2)

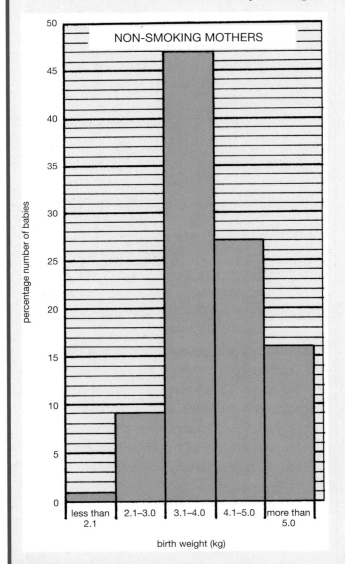

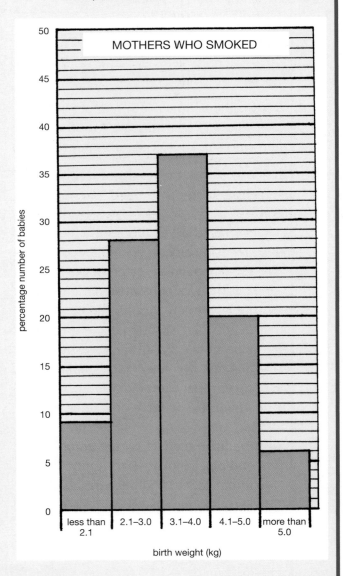

**Figure 3.36**

## What you should know

| | | |
|---|---|---|
| asthma | deeper | peak |
| baby | disease | pregnant |
| blood | exchange | recovery |
| breath | fitness | sacs |
| bronchioles | heart | shorter |
| bronchus | increases | size |
| cancer | lungs | tar |
| capillaries | maximum | tidal |
| carbon dioxide | meter | volume |
| carbon monoxide | oxygen | windpipe |

**Table 3.12**  *Word bank for chapter 3*

1  The lungs are situated in the chest. Air enters the _____ by passing down a tube called the _____ which divides into two branches. Each branch, called a _____, divides into many smaller tubes called _____. Each bronchiole ends in several air _____. Each air sac is surrounded by tiny blood vessels called _____.

2  The function of the lungs is to take _____ from the air into the blood and to remove _____ from the blood. This exchange of gases takes place between the air sacs and the _____ in the capillaries.

3  Breathing rate is measured by counting the number of breaths taken per minute. Exercise _____ breathing rate and makes the person take _____ breaths. Both these effects increase the rate of gas _____ in the lungs.

4  During _____ time after exercise, rate and depth of breathing return to normal. The _____ the recovery time, the fitter the person.

5  _____ volume is the volume of air breathed in or out of the lungs in one normal breath. It can be measured using a breath _____ kit.

6  Vital capacity is the _____ volume of air that can be breathed out in one _____ after a maximum inspiration. It can be measured using a tank of water, a bell jar and a tube.

7  _____ flow is the maximum rate at which air can be forced from the lungs. Peak flow can be measured using a peak flow _____. It is used in diagnosis and management of the respiratory condition _____.

8  Tidal volume, vital capacity and peak flow vary from person to person and depend on factors such as body _____, age, sex and level of _____.

9  Cigarette smoking seriously damages a person's health. The _____ in the smoke increases the risk of lung _____.

10  A gas called _____ in cigarette smoke reduces the blood's ability to carry oxygen round the body. The _____ has to work harder and this can lead eventually to heart _____. Smoking when _____ can also damage the health of the unborn _____.

## 4  A healthy body

## Food

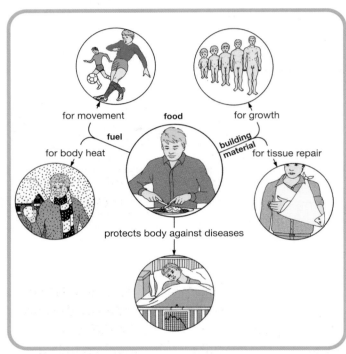

**Figure 4.1** *The body needs food for many purposes*

Figure 4.1 shows how food is needed by the body to:

- provide it with fuel for **energy**;
- supply it with **building materials** for growth and tissue repair;
- **protect** it against diseases.

### Food groups and their uses

Foods can be divided into different groups based on the use that the body makes of them (see figure 4.2). Foods rich in fat or carbohydrate give the body energy. Foods rich in protein are used by the body for growth and tissue repair. Foods rich in vitamins and minerals protect the body against deficiency diseases (see tables 4.1 and 4.2).

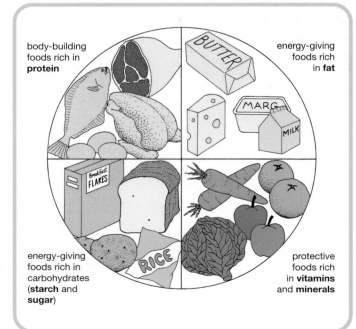

**Figure 4.2** *Food groups*

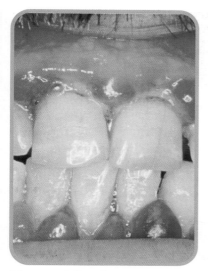

**Figure 4.3** *Scurvy causing bleeding gums*

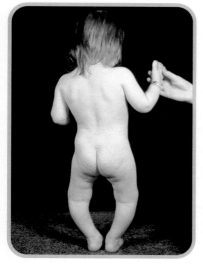

**Figure 4.4** *This person is suffering from rickets*

| vitamin | deficiency disease resulting from shortage of vitamin | rich source of vitamin |
|---|---|---|
| A | night-blindness | carrots |
| C | scurvy (poor healing of wounds; soft, bleeding gums) – see figure 4.3 | oranges |
| D | rickets (soft bones that become deformed easily) – see figure 4.4 | whole milk |

**Table 4.1** *Importance of vitamins*

| mineral | result of serious deficiency of mineral | rich source of mineral |
|---|---|---|
| calcium | weak bones and soft teeth | milk |
| iron | anaemia (shortage of red blood cells) | red meat, green vegetables |

**Table 4.2** *Importance of minerals*

## Importance of a balanced diet

The food and drink that a person normally consumes is called their **diet**. No single food contains all of the chemicals needed by the body for good health. A balanced mixture of foods from the different groups is needed to make up a healthy diet. This should include at least five portions of fruit and vegetables every day. An example of a **balanced diet** is shown in figure 4.5.

**Figure 4.5** *A balanced diet*

67

## Energy

Energy is needed by the body for **growth**, **movement**, **keeping warm**, **heartbeat** and many other workings of the body. All of this energy comes from **food**.

### Activity

## Comparing the energy released by two foods

**You need**

- clamp stand
- boiling tube
- measuring cylinder (100 cm³)
- mounted needle
- peanut (or almond nut)
- Bunsen burner
- thermometer
- piece of macaroon bar

| food | temperature of water (°C) | | rise in temperature (°C) |
| --- | --- | --- | --- |
| | at start | after food has burned | |
| peanut/ almond nut | | | |
| macaroon bar | | | |

**Table 4.3**   *Energy release results*

**What to do**

**1** Prepare a copy of table 4.3.

**2** Follow the procedure illustrated in figure 4.6.

**3** Repeat the experiment using 1 g of macaroon bar (a food rich in starch and sugar).

**4** Draw a conclusion about the quantity of energy released by the two foods.

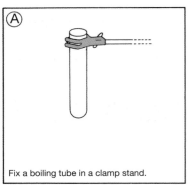
Ⓐ Fix a boiling tube in a clamp stand.

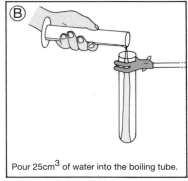
Ⓑ Pour 25cm³ of water into the boiling tube.

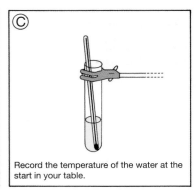
Ⓒ Record the temperature of the water at the start in your table.

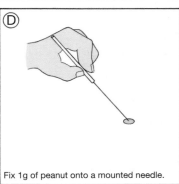
Ⓓ Fix 1g of peanut onto a mounted needle.

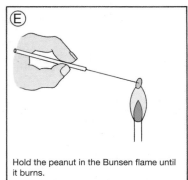
Ⓔ Hold the peanut in the Bunsen flame until it burns.

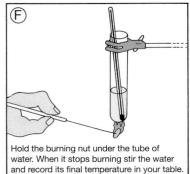
Ⓕ Hold the burning nut under the tube of water. When it stops burning stir the water and record its final temperature in your table.

**Figure 4.6**   *Measuring the energy released by a burning food*

## Energy content of foods

Energy is measured in units called kilojoules (kJ). Different foods have different energy values. Some examples are shown in table 4.4.

| food | energy value per 100 g portion (kJ) |
|------|--------------------------------------|
| beans | 390 |
| beef | 1360 |
| biscuit | 2350 |
| boiled potatoes | 370 |
| bread | 1010 |
| cheese | 1780 |
| chicken | 580 |
| chips | 1000 |
| egg | 680 |
| margarine | 3500 |
| milk | 280 |
| peanut | 2550 |

**Table 4.4** *Energy values for foods*

## Energy requirements

| person | daily energy requirement (kJ) |
|--------|-------------------------------|
| 2 year-old child | 5000 |
| 6 year-old child | 6500 |
| 12–15 year-old girl | 9600 |
| 12–15 year-old boy | 11 700 |
| woman (light work) | 9500 |
| woman (pregnant) | 10 000 |
| woman (heavy work) | 12 500 |
| man (light work) | 11 500 |
| man (moderate work) | 13 000 |
| man (very heavy work) | 15 500 |

**Table 4.5** *Daily energy requirements*

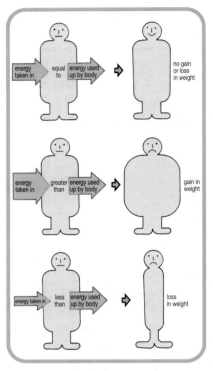

**Figure 4.8** *Energy balance and imbalance*

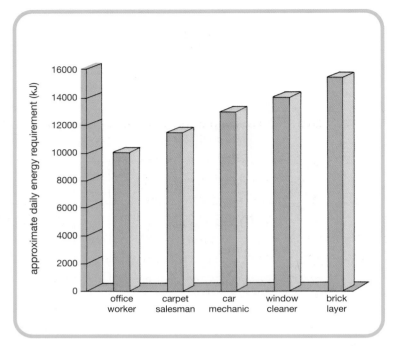

**Figure 4.7** *Energy requirements of different occupations*

The total amount of energy required per day depends on several factors such as the person's:

- **age** (adults and growing teenagers normally need more energy than young children – see table 4.5);
- **sex (gender)** (men normally need more energy than women – see table 4.5);
- **body size** (people with a bigger body frame need more energy than those with a smaller build);
- **occupation** (people doing heavy work need more energy than those doing lighter work – figure 4.7);
- **leisure pursuits** (some sports and activities use up more energy than others – see table 4.6).

## Energy balance and imbalance

When the energy contained in the food that a person takes in is equal to the energy used by the body, the person is in a state of energy balance and neither gains nor loses weight. However if the person takes in more energy than their body uses up, they suffer an energy imbalance. The extra energy is stored as fat and they gain weight. On the other hand, if the person takes in less energy than their body needs, they lose weight (see figure 4.8).

| leisure activity | approximate energy used per hour (kJ) |
| --- | --- |
| watching TV | 400 |
| walking | 1000 |
| swimming | 1500 |
| tennis | 1700 |
| football | 2150 |
| running | 2500 |
| boxing | 3200 |

**Table 4.6** *Energy requirements of different leisure activities*

## Activity

### Analysing diets

*What to do*

Study table 4.7 which gives the typical diet for one day of four pupils in a Scottish school. Answer the questions that follow.

**1** a) Whose diet is unbalanced because it contains excessive amounts of energy-giving foods?
   b) Predict what will happen to this unnecessary fuel.
**2** Whose diet is balanced and contains five or more portions of fruit and vegetables?
**3** Whose diet is balanced except that it does not contain enough fruit and vegetables?
**4** Whose diet is unbalanced because it does not contain enough energy-giving foods?
**5** Describe the likely appearance of a) Stephen, b) Nicola.

| | **Alan** | **Nicola** | **Stephen** | **Jane** |
|---|---|---|---|---|
| **breakfast** | 1 bowl of Alpen half pint of milk 1 slice of toast and margarine 1 boiled egg 1 cup of tea (+ milk and sugar) | 1 cup of tea (+ skimmed milk) | 1 large bowl of Coco Pops half pint of milk 2 slices of toast and butter 1 bar of chocolate 1 can of cola | 1 Weetabix and sliced banana half pint of semi-skimmed milk 1 slice of toast and margarine 1 cup of tea (+ semi-skimmed milk) |
| **lunch** | steak pie peas mashed potato sponge and custard | 1 can of Diet Coke 1 bag of cheese and onion crisps | 1 hot dog chips 1 ice cream 1 chocolate flake 1 can of cola 1 bag of salted nuts | mince cabbage jacket potato 1 orange 1 pear |
| **evening meal** | mushroom soup roast chicken chips turnip 1 pear | chicken salad 1 cup of tea (+ skimmed milk) | 2 fried hamburgers chips beans fresh cream trifle 2 chocolate biscuits 2 cans of Irn Bru | grilled fish chips broccoli 1 fruit yoghurt 1 apple 1 can of diet cola |

**Table 4.7** *Analysing four diets*

## Body fat and health

The human body needs fat for a variety of reasons.

- Fat acts as an **energy store** and releases twice as much energy as the same weight of carbohydrate.
- Fat stored under the skin **insulates** the body by cutting down heat loss.
- Fat pads in the hands and feet **absorb shock**.
- Fat is needed to make **nerve cells** work.

For good health about 15–20% of an adult male's body mass and about 20–25% of an adult female's body mass should consist of fat (see figure 4.9).

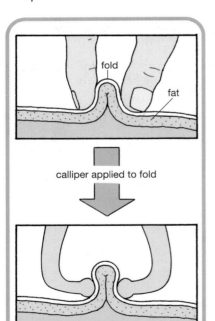

**Figure 4.10** *A skin fold calliper*

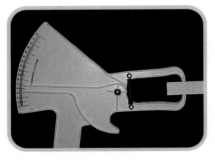

calliper applied to fold

**Figure 4.11** *Taking a skin fold measurement*

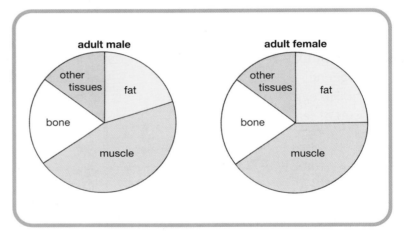

**Figure 4.9** *Percentage of fat and other tissues in humans*

### Use of a skin fold calliper

A **skin fold calliper** (see figure 4.10) is a low-tech instrument used to measure the thickness of a fold of skin and the layer of fat that the fold contains. A fold of skin is taken between the thumb and index finger and the calliper is applied to the fold (see figure 4.11).

Skin fold measurements are normally taken at the four key positions shown in figure 4.12 on the right-hand side of the person's body. Although the diagram shows bare skin, there is no need to remove clothing as long as its thickness is subtracted from the measurements taken.

Body fat can also be measured using a fat **sensor** (see page 6). This high-tech instrument resembles a set of bathroom scales and gives the person's fat content as a digital display.

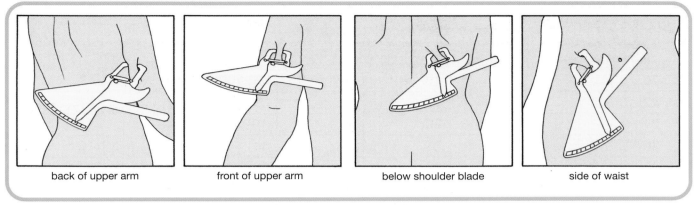

back of upper arm        front of upper arm        below shoulder blade        side of waist

**Figure 4.12**  *Key positions used for taking skin fold measurements*

Activity

## Measuring body fat using a skin fold calliper

*You need*
- skin fold calliper

*What to do*
1 Prepare a copy of tables 4.8 and 4.9.
2 Use the calliper to take three skin fold measurements for each of the four locations shown in figure 4.10 and complete table 4.8 as you go along.
3 Calculate the four averages and add them up to give the sum of the skin fold measurements and enter this in table 4.9.
4 Make a correction to allow for layers of clothing trapped in the calliper and subtract this from the total.
5 Refer to table 4.10 to find out the percentage fat content of the person's body.

| region of body | skin fold calliper reading (mm) | | | |
|---|---|---|---|---|
| | 1 | 2 | 3 | average |
| back of upper arm | | | | |
| front of upper arm | | | | |
| below shoulder blade | | | | |
| side of waist | | | | |

**Table 4.8**  *Results for skin fold calliper*

| | |
|---|---|
| **sum of average skin fold measurements and clothing (mm)** | |
| **correction for clothing (mm)** | |
| **sum of average skin fold measurements without clothing (mm)** | |
| **fat content of body (%)** | |

**Table 4.9**  *Totalled and corrected results for skin fold calliper*

| sum of average skin fold measurements (mm) | fat content of body (%) | | | | | |
|---|---|---|---|---|---|---|
| | woman's age (years) | | | man's age (years) | | |
| | 16–29 | 30–49 | 50 & over | 16–29 | 30–49 | 50 & over |
| 20 | 14 | 18 | 21 | 8 | 12 | 13 |
| 22 | 15 | 19 | 23 | 9 | 13 | 14 |
| 24 | 17 | 20 | 24 | 10 | 14 | 15 |
| 26 | 18 | 21 | 25 | 11 | 15 | 16 |
| 28 | 19 | 22 | 26 | 12 | 16 | 17 |
| 30 | 20 | 23 | 27 | 13 | 17 | 19 |
| 35 | 22 | 25 | 29 | 15 | 19 | 21 |
| 40 | 23 | 27 | 30 | 16 | 20 | 23 |
| 45 | 25 | 28 | 32 | 18 | 22 | 25 |
| 50 | 27 | 30 | 33 | 19 | 23 | 26 |
| 55 | 28 | 31 | 35 | 20 | 24 | 28 |
| 60 | 29 | 32 | 36 | 21 | 25 | 29 |
| 65 | 30 | 33 | 37 | 22 | 26 | 30 |
| 70 | 31 | 34 | 38 | 23 | 27 | 32 |
| 75 | 32 | 35 | 39 | 24 | 28 | 33 |
| 80 | 33 | 36 | 40 | 25 | 29 | 34 |
| 85 | 34 | 36 | 40 | 26 | 30 | 35 |
| 90 | 35 | 37 | 41 | 26 | 30 | 36 |
| 95 | 36 | 38 | 42 | 27 | 31 | 37 |
| 100 | 37 | 39 | 43 | 28 | 32 | 38 |
| 110 | 38 | 40 | 44 | 29 | 33 | 39 |
| 120 | 39 | 41 | 45 | 30 | 34 | 40 |
| 130 | 40 | 42 | 46 | 31 | 35 | 42 |
| 140 | 41 | 43 | 47 | 32 | 36 | 43 |
| 150 | 42 | 44 | 48 | 33 | 37 | 44 |
| 160 | 43 | 45 | 49 | 34 | 38 | 45 |
| 170 | 44 | 46 | 50 | 34 | 39 | 46 |
| 180 | 45 | 46 | 51 | 35 | 39 | 47 |
| 190 | 46 | 47 | 52 | 36 | 40 | 48 |
| 200 | 47 | 48 | 52 | 37 | 41 | 49 |

**Table 4.10** *Percentage fat content of human body*

# Range of values of body mass

There is an ideal body mass (weight) for everyone. This varies from person to person depending on their height, age and sex.

## Body mass index (BMI)

A person's BMI is calculated by using the formula:

$$BMI = \frac{body\ mass}{height^2}$$

(where body mass is measured in kg and height is measured in m).

BMI is commonly used to find out if a person's body mass is in the ideal range for their height (see table 4.11).

| BMI value | opinion of experts |
|-----------|--------------------|
| below 20 | underweight |
| 20–25 | ideal for height |
| 26–30 | overweight |
| 31–35 | obese (very overweight) |
| over 35 | very obese (grossly overweight) |

**Table 4.11**   *BMI values*

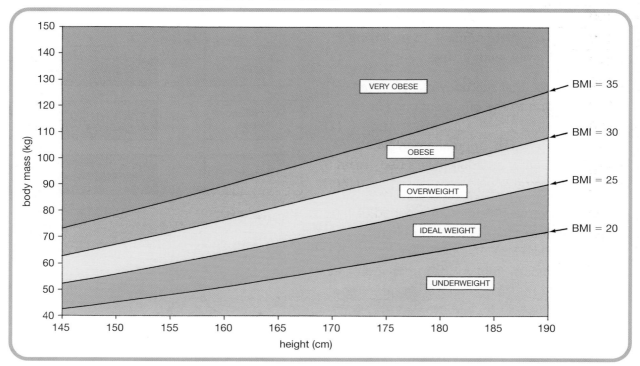

**Figure 4.13**   *BMI graph*

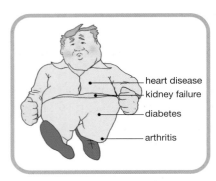

**Figure 4.14** *Obesity and health problems*

Another way of doing this is to use the BMI graph shown in figure 4.13. A straight line is drawn across from the person's body mass and up from their height. The point where the two lines meet gives the BMI value.

It should be remembered that these measurements are not very reliable for young people because many young people store fat during adolescence and lose it later when they stop growing. These BMI values are only reliable for adults who have stopped growing.

## Implications for health

When the quantity of energy in food eaten is regularly greater than the quantity of energy used up by the body, the person gains weight. People who are more than 15% heavier than the ideal body mass for their height are said to be obese. Being overweight can lead to an increased risk of the health problems shown in figure 4.14.

The person should lose weight by cutting down on food rich in fat and sugar and by increasing their intake of fruit and vegetables.

## Underweight

Being significantly underweight is rare. However when severe weight loss occurs over a period of months or years, it can indicate serious problems such as anorexia or cancer.

## Testing your knowledge

1 Copy and complete figure 4.15 by linking each food group with its use. (3)

2 a) Give a use that the body makes of fat in addition to the one mentioned in question 1. (1)
   b) For good health, approximately what percentage of an adult male's body mass should be fat? (1)

3 a) Being overweight for a long period of time can lead to serious health problems. Name THREE of these. (3)
   b) Being underweight for a long period of time can indicate that the person is suffering a serious health problem. Name TWO such problems. (2)

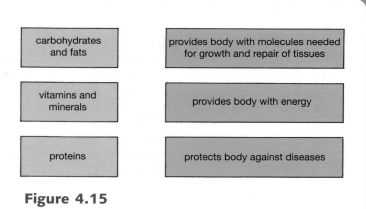

**Figure 4.15**

# Body temperature and health

The **normal temperature** of the human body is 37°C. The body can only stay healthy and work properly if it is kept fairly close to this temperature. The **normal range** of body temperature is 36.0–37.5°C. Body temperature can be measured by using a variety of types of **thermometer**.

## Activity

## Measuring body temperature using a glass clinical thermometer (assessment)

### Information

A **glass clinical thermometer** is a low-tech instrument. It normally contains mercury (or coloured liquid) in a bulb (see figure 4.16). The bend in the thread stops the mercury rushing back down the tube when the thermometer bulb is removed from the patient's mouth. This allows an accurate reading to be taken. The thermometer needs to be shaken to return the mercury to the bulb.

### You need
- glass clinical thermometer
- disinfectant
- paper towels
- stopwatch

| reading | body temperature (°C) |
|---------|------------------------|
| 1 | |
| 2 | |
| average | |

**Table 4.12**  *Glass clinical thermometer results*

### What to do
**1** Prepare a copy of table 4.12.
**2** Carefully take the thermometer out of its case. Practise reading it by holding the thermometer horizontally with the bulb to the left and turning the thermometer round until the mercury column can be seen on the scale.
**3** If the mercury level is above 35°C, shake the thermometer until the mercury is below 35°C. (This is done by holding the thermometer by the end opposite to the bulb and flicking the thermometer as if cracking a whip.)
**4** Clean the bulb end with disinfectant and dry it on a paper towel.
**5** Place the bulb in your armpit and start the stopwatch. (*Note*: The bulb of this type of thermometer is normally placed under the patient's tongue when used by a doctor.)
**6** After two minutes take the thermometer out of your armpit, read the temperature and record the result in the table.
**7** Repeat steps 3, 5 and 6 and complete the table. (*Note*: Taking a second reading and calculating an *average* gives an overall result that is more *reliable*.)
**8** Ask the teacher to check your second reading and assess your work.
**9** Clean the thermometer bulb with disinfectant and return it to its case.

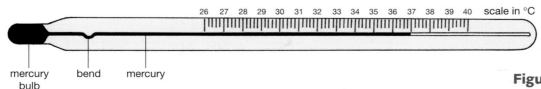

26  27  28  29  30  31  32  33  34  35  36  37  38  39  40   scale in °C

mercury   bend   mercury
bulb

**Figure 4.16**  *Glass clinical thermometer*

## Activity

## Measuring body temperature using a digital clinical thermometer

### Information

A **digital clinical thermometer** (see figure 4.17) is a high-tech instrument often used in place of a glass clinical thermometer.

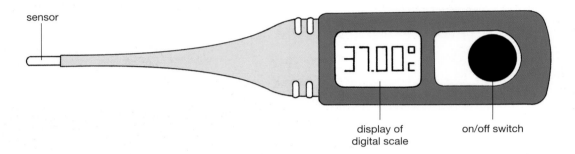

sensor

37.00 °C

display of digital scale

on/off switch

**Figure 4.17** *Digital clinical thermometer*

### You need

- digital clinical thermometer
- disinfectant
- paper towels

|  |  | region of the body | |
|---|---|---|---|
|  |  | **mouth** | **clenched fist** |
| **temperature (°C)** | first reading |  |  |
|  | second reading |  |  |
|  | average |  |  |

**Table 4.13** *Digital clinical thermometer results*

### What to do

1 Prepare a copy of table 4.13.
2 Clean the sensor end of the thermometer with disinfectant and dry it on a paper towel.
3 Place the sensor in the person's mouth well under their tongue.
4 Press the thermometer's on/off switch once.
5 Get the person to keep their mouth closed and to hold the thermometer steady while the digits change.
6 Read the person's body temperature once the reading has been stable on the display panel for 10 seconds and record it in the table.
7 Press the on/off switch again to turn the display off.
8 Repeat steps 3–6 to take a second reading for the mouth.
9 Press the on/off switch again to turn the display off.
10 Repeat the procedure to take two readings of temperature for the same person's clenched fist.
11 Clean the sensor with disinfectant after use.
12 Complete the table and draw a conclusion from your results.
13 Answer the following questions:
  a) Which result is a measure of the internal body temperature?
  b) Which result is a measure of the surface body temperature?
  c) Why is it good scientific practice to take a second reading each time and calculate an average?

## Activity

# Measuring body temperature using a liquid crystal strip thermometer

### Information

A **liquid crystal strip thermometer** (see figures 4.18 and 4.19) is a high-tech instrument made of strong flexible plastic. It allows surface body temperature measurements to be made on different parts of the body. When the strip is held against the skin, the squares start to 'light up'. After 10 seconds, the square that is lit up 'green' indicates the correct temperature.

### You need

- liquid crystal thermometer
- disinfectant
- soft paper tissue
- calculator

### What to do

**1** Prepare a copy of table 4.14.

**2** Clean the thermometer by gently wiping it with a tissue moistened with disinfectant.

**Figure 4.18** *Liquid crystal strip thermometer*

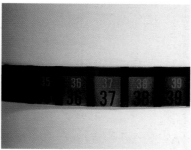

**Figure 4.19** *Close-up of 'lit-up' box*

**3** Get your partner to help you to hold the thermometer flat against the palm of your hand.

**4** After 10 seconds, record the temperature in the table and do two more readings for other parts of the palm of the hand.

**5** Repeat steps 3–4 for the other regions of the body in the table.

**6** Clean the thermometer.

**7** Complete the table by calculating averages.

**8** Draw a conclusion from your results.

**9** a) Explain why it is good scientific practice to calculate an average result.

b) Suggest why a doctor uses a clinical and *not* a liquid strip crystal thermometer to measure temperature.

| region of body | temperature (°C) | | | |
|---|---|---|---|---|
| | first reading | second reading | third reading | average |
| palm of hand | | | | |
| wrist | | | | |
| forehead | | | | |
| neck | | | | |

**Table 4.14** *Liquid crystal strip thermometer*

## Effects of high and low body temperatures

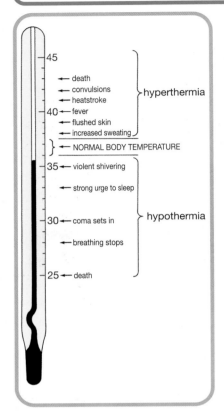

**Figure 4.20** *Extremes of body temperature*

Figure 4.20 indicates what happens to the body at body temperatures above and below the normal range of 36.0–37.5°C.

### Increase in body temperature

#### Heatstroke

When a person is exposed to intense heat and high humidity, they are unable to cool down by sweating and their body temperature increases. If their temperature rises above 40°C, then they suffer **heatstroke** and urgently need medical treatment to cool them down. In extreme cases, heatstroke results in *convulsions, coma* and *death*.

#### Fever

When the body burns its energy reserves to fight off an infection, the heat generated causes the body temperature to rise dramatically. This unusually high temperature (e.g. 40°C or more) is known as a **fever** and the person urgently needs medical treatment.

### Decrease in body temperature

When a person is exposed to a cold environment for a long time but is unable to keep warm by shivering, drinking hot fluids etc., their body temperature decreases. If it drops below 35°C, then they suffer **hypothermia** and urgently need medical aid to warm them up again. In extreme cases, hypothermia results in *coma* and *death*.

**Babies** and **elderly people** are particularly at risk from hypothermia because their bodies are not good at shivering and generating the heat needed to warm up the body.

## Exercise and health of muscles

Much of the energy gained from food is used for movement. Movement of the human body is brought about by the action of more than six hundred **muscles**. A few are shown in figure 4.21.

Figure 4.22 shows the action of two arm muscles. When the **biceps** muscle contracts, the arm bends. When the **triceps** muscle contracts, the arm straightens. Muscles need to be exercised regularly to maintain their size and strength. Regular workouts make muscles develop an improved **blood supply**, **grow bigger** and increase in **strength**. If muscles are not exercised regularly, they decrease in size and strength.

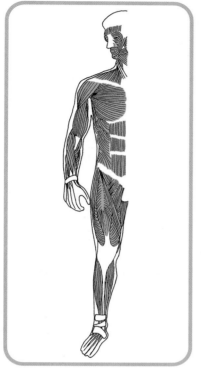

**Figure 4.21** *Muscles*

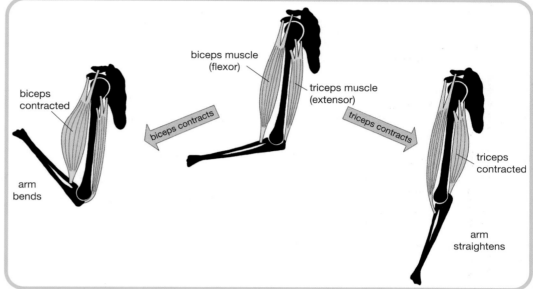

biceps muscle (flexor)

triceps muscle (extensor)

biceps contracted

biceps contracts

arm bends

triceps contracts

triceps contracted

arm straightens

**Figure 4.22** *Action of two arm muscles*

## Activity

## Measuring muscle strength

### Information

**Muscle strength** is a measure of the force that can be exerted by muscles. The strength of the muscles used in a handgrip can be measured using a **grip strength dynamometer** (see figure 4.23) or a set of **bathroom scales** (see figure 4.24).

### You need

- grip strength dynamometer

### What to do

1 Prepare a copy of table 4.15.
2 Adjust the dynamometer to suit your hand size.
3 Set the scale to zero.

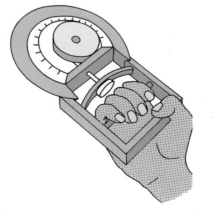

**Figure 4.23** *Dynamometer*     **Figure 4.24** *Bathroom scales*

**4** While holding the dynamometer in your right hand, extend your right arm down your side but clear of your body.

**5** Squeeze the handles *as hard as you can* for two seconds.

**6** Ask your partner to enter the result in your table.

**7** Repeat steps 3–6 twice.

**8** Complete the table by entering your best result as a measure of your handgrip strength.

| attempt number | handgrip strength (N) |
|---|---|
| 1 | |
| 2 | |
| 3 | |
| best of 3 | |

**Table 4.15** *Muscle strength results*

## Investigation

## Investigating whether the left hand or the right hand is stronger (assessment)

### Information

Hand grip strength can be measures in **newtons (N)** using a **grip strength dynamometer** (see figure 4.25). Your job is to plan and carry out an investigation to find out whether the left hand or the right hand has the greater handgrip strength.

### What to do

**1** Make up your plan by answering the following questions:
  a) What is the aim of this investigation?
  b) What factor are you going to measure?
  c) What factor are you going to alter?
  d) Name one factor that will be kept the same throughout the investigation.
  e) How many attempts will you make each time?
  f) What apparatus will you use?
  g) How will you record your results?

**2** Ask your teacher to check your plan before you carry out the investigation.

**3** Prepare your table of results.

**4** Carry out the investigation with the help of your partner and complete the results table as you go along.

**5** Make a bar chart of your best value for each hand.

**6** Draw a conclusion from your results.

**7** Identify a strength or a weakness of your investigation.

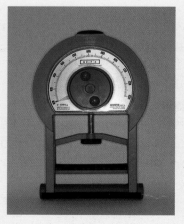

**Figure 4.25** *Grip strength dynamometer*

## Measuring muscle fatigue

### Information

Muscles need **oxygen** to work. During exercise, uptake of oxygen by muscles increases. When muscles are exercised non-stop, there comes a point when they run out of the oxygen needed for energy release. When this happens, they become **fatigued** (tired) and work less efficiently or may even completely refuse to work. A period of rest and relaxation relieves this muscle fatigue by allowing time for fresh supplies of oxygen to reach the muscles.

### You need

- grip strength dynamometer

### What to do

1 Prepare a copy of table 4.16.
2 Adjust the dynamometer to suit your hand size.
3 Set the scale to zero.
4 While holding the dynamometer in your right hand, extend your right arm down your side but clear of your body.
5 Squeeze the handles *as hard as you can* for two seconds.
6 Ask your partner to enter the result in your table.
7 Immediately repeat steps 3–6 and keep doing so until you have made 10 attempts without taking any rests between them.
8 Draw a line graph of the results with attempt number on the horizontal (x) axis.
9 Answer the following questions:
   a) (i) What was your highest value for handgrip strength in this experiment?
   (ii) Which attempt produced this result?
   b) (i) Between which two attempts did muscle fatigue begin to set in?
   (ii) How could you tell from your results?
   c) Which attempt was affected most by muscle fatigue?
   d) Predict whether handgrip strength after a rest of several minutes would be greater, equal to or less than that found at attempt 10. (If possible, try it and see if you were right.)

| attempt number | handgrip strength (N) |
|---|---|
| 1 | |
| 2 | |
| 3 | |
| 4 | |
| 5 | |
| 6 | |
| 7 | |
| 8 | |
| 9 | |
| 10 | |

**Table 4.16** *Muscle fatigue results*

## Testing your knowledge

1 Copy figure 4.26 and complete it by using arrows to connect each term with its temperature. (3)

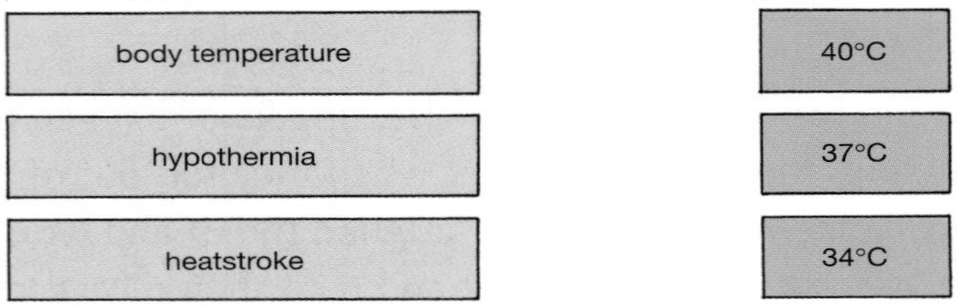

| body temperature | | 40°C |
| hypothermia | | 37°C |
| heatstroke | | 34°C |

**Figure 4.26**

2 a) Name a high-tech and a low-tech method of measuring body temperature. (2)

b) Why does body temperature rise when a person is suffering from a disease such as yellow fever? (1)

3 Give TWO signs that indicate that an elderly person is suffering hypothermia. (2)

4 Decide whether each of the following statements is true or false and then use T or F to indicate your choice. Where a statement is false, give the word that should have been used in place of the word in **bold** print. (5)

a) Regular exercise maintains the size and **strength** of muscles.

b) During exercise, **carbon dioxide** uptake by muscles increases.

c) Muscle strength can be measured using a **sphygmomanometer**.

d) Muscle **fatigue** occurs during exercise if the muscles do not get enough oxygen.

e) The size and strength of a pair of muscles **increase** if they are not exercised regularly.

# Reaction time

In many sporting activities, it is important to react quickly to a certain signal. The time taken to respond to a stimulus (e.g. the starting gun) is called reaction time. It can be measured using a dropped metre stick (see below) or an electronic timer.

Reaction time can be affected by drugs, alcohol or excitement. Reaction time is a useful indicator of a person's state of health. A long (slow) reaction time can indicate that the person is suffering one or more of the following:

- diabetes;
- brain disorder;
- nerve disorder;
- arterial disease.

# Measuring reaction time using a dropped metre stick (low-tech method)

**You need**
- metre stick

**What to do**

1 Read all of the following instructions and then prepare a table to record results for yourself and your partner before beginning.

2 Hold the metre stick just above your partner's hand (see figure 4.27 part A).

3 Let the metre stick go unexpectedly. Your partner must grab it as quickly as possible.

4 Note the length in centimetres that has fallen through their hand (see figure 4.27 part B).

5 Refer to table 4.17 to find out the reaction time and enter it in your table.

6 Repeat the procedure four times and add the results to your table.

7 Repeat the experiment with your partner in control of the metre stick and you reacting five times to it being dropped.

8 Complete your table of results.

9 Compare results with your partner and other people in the class.

10 Answer the following questions:
   a) In general what happens to reaction time with practice?
   b) Explain why practice cannot improve reaction time beyond a certain limit.

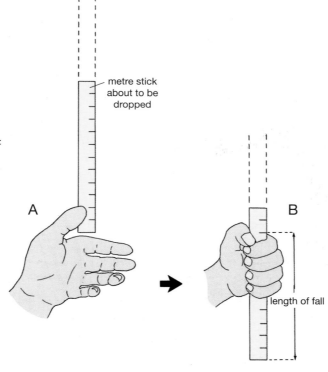

**Figure 4.27** *Measuring reaction time*

| length (cm) | reaction time (seconds) |
|---|---|
| 1 | 0.045 |
| 2 | 0.064 |
| 3 | 0.078 |
| 4 | 0.090 |
| 5 | 0.101 |
| 6 | 0.111 |
| 7 | 0.120 |
| 8 | 0.128 |
| 9 | 0.136 |
| 10 | 0.143 |
| 11 | 0.150 |
| 12 | 0.156 |
| 13 | 0.163 |
| 14 | 0.169 |
| 15 | 0.175 |
| 16 | 0.181 |
| 17 | 0.186 |
| 18 | 0.192 |
| 19 | 0.197 |
| 20 | 0.202 |
| 21 | 0.207 |
| 22 | 0.212 |
| 23 | 0.217 |
| 24 | 0.221 |
| 25 | 0.226 |
| 26 | 0.230 |
| 27 | 0.235 |
| 28 | 0.239 |
| 29 | 0.243 |
| 30 | 0.247 |
| 31 | 0.252 |
| 32 | 0.256 |
| 33 | 0.260 |
| 34 | 0.263 |
| 35 | 0.267 |
| 36 | 0.271 |
| 37 | 0.275 |
| 38 | 0.278 |
| 39 | 0.282 |
| 40 | 0.286 |

**Table 4.17** *Reaction times*

# Health risks of alcohol

## Alcohol

A **drug** is a substance that alters the way the human body works. **Alcohol** is the most commonly used drug in Britain. Alcoholic drinks vary in the percentage of alcohol that they contain as shown in figure 4.28.

Once consumed, alcohol is absorbed from the stomach and other parts of the gut and carried round the body in the bloodstream (see figure 4.29). When it reaches the **brain**, it affects the brain's activity. This often makes the person feel relaxed and uninhibited. The alcohol is slowly broken down into harmless substances by the **liver** and the effect of the alcohol gradually wears off.

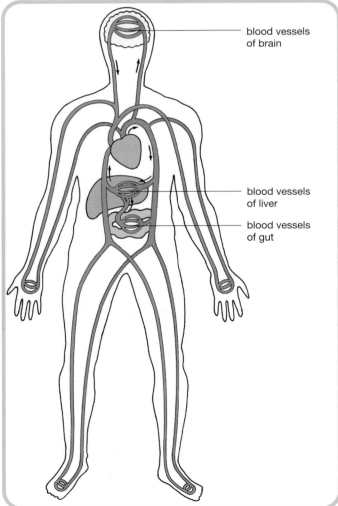

blood vessels of brain

blood vessels of liver

blood vessels of gut

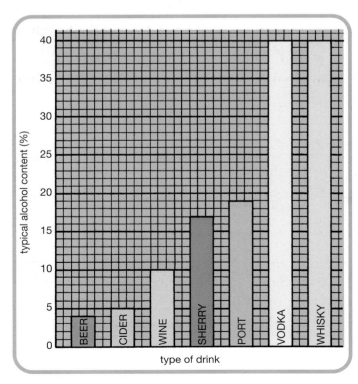

**Figure 4.28** *Alcohol content of different drinks*

**Figure 4.29** *Route taken by alcohol around the body*

**Figure 4.30** *Examples of one unit of alcohol*

### One unit

One unit of alcohol is the quantity that can be broken down by a healthy adult's liver in about one hour. Figure 4.30 shows three different drinks that each contain about one unit of alcohol. The recommended upper limit per week for women is 14 units and for men is 21 units. It has been shown that drinking more than this can cause damage to the body.

### Harmful effects

If a large quantity of alcohol is consumed, the liver does not have time to break it down before the alcohol has caused the following harmful effects.

In the **short term**, excess alcohol in the bloodstream leads to:

- **longer** (slower) **reaction time**;
- **poorer muscle control** and coordination;
- **poorer judgement** when faced with a decision to make.

All of these effects increase the risk of an accident taking place. In the UK, drink driving is responsible for:

- around 1000 people being killed annually;
- road accidents being the largest cause of death amongst young people;
- the death of 1 in every 4 drivers killed on the road.

Women who take alcohol during **pregnancy** can damage the health of the unborn baby.

In the **long term**, drinking alcohol to excess on a regular basis leads to **liver** and **brain** damage.

## Measurement of alcohol in exhaled air

### Breathalyser

This piece of low-tech equipment is shown in figure 4.31. The person suspected of being 'over the limit' blows into the mouthpiece. Alcohol makes the crystals change in colour from yellow to green. If the crystals go green beyond the centre line, the person has failed the test.

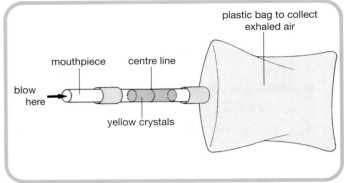

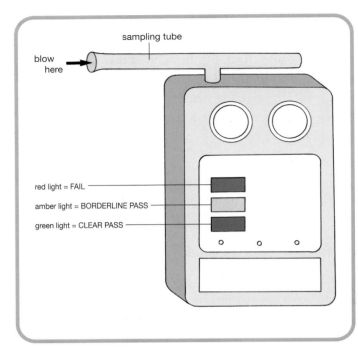

**Figure 4.32**   *An alcometer*

**Figure 4.31**   *A breathalyser*

## Alcometer

This piece of high-tech equipment is shown in figure 4.32. The person blows into the sampling tube. An electronic sensor picks up the alcohol present in the exhaled air. In this type of **alcometer** the result is indicated by a coloured light (see diagram). In some alcometers, the result is given as a digital display of blood alcohol content.

Neither of the pieces of equipment shown in figures 4.31 and 4.32 produces results that are accurate enough to be used as evidence in court. A driver who fails a roadside test is arrested and taken to a police station to allow more accurate tests to be carried out.

## Activity

### Measuring the alcohol content of 'exhaled' air

#### Information
This activity is intended to be carried out as a demonstration by the teacher

#### Teacher needs
- sterile plastic bag (approximately 30 cm²)
- length of glass/plastic tubing (approximately 10 cm)
- elastic band
- alcometer and sampling tube
- short length of rubber tubing to fit to alcometer's sampling tube
- plastic dropper
- alcohol
- hair drier

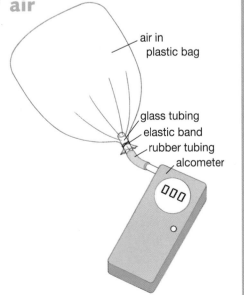

**Figure 4.33**   *Measuring the alcohol content of air*

## What to do

1 Watch as the teacher:
- inflates the plastic bag with air;
- encloses the bag's open end around the plastic tube using the elastic band;
- fits the glass tube to the alcometer (see figure 4.33);
- turns on the alcometer;
- squeezes the air into the alcometer and notes the result;
- detaches the plastic bag and uses the dropper to squirt a small volume (e.g. 2 cm³) of alcohol into it;
- uses the hair drier to blow warm air into the bag for two minutes to promote evaporation of the alcohol;
- repeats the first five steps in the procedure.

2 Make a simple diagram of the apparatus set up and in use.

3 Construct a table of results.

4 Draw a conclusion from the results.

## Testing your knowledge

1 a) What is meant by the term *reaction time*? (1)
   b) What effect do the following factors have on reaction time?
      (i)   practice,
      (ii) drugs. (2)
   c) A long reaction time can indicate that the body may be suffering from a disorder. Name TWO such disorders. (2)

2 a) By what means does alcohol get from the stomach to the brain? (1)
   b) What effect does alcohol have on
      (i)   the drinker's brain?
      (ii)  the drinker's behaviour? (2)

3 a) What is meant by the term *one unit* of alcohol? (1)
   b) What is the recommended upper limit (in units per week) for
      (i)   women?
      (ii)  men? (2)
   c) Give TWO short-term harmful effects of alcohol. (2)
   d) Give TWO long-term harmful effects of alcohol. (2)

## Applying your knowledge

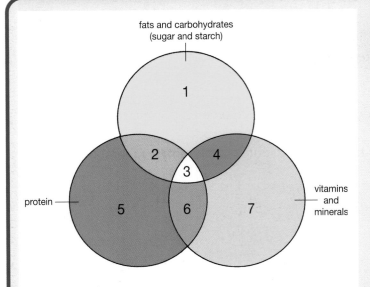

fats and carbohydrates
(sugar and starch)

protein

vitamins and minerals

**Figure 4.34**

| food | main food types present | number on Venn diagram |
|---|---|---|
| a) streaky bacon | protein and fat | |
| b) oranges | sugar and vitamin C | |
| c) milk | sugar, protein, fat and many vitamins and minerals | |
| d) butter | fat | |
| e) fish | proteins and minerals | |
| f) egg white | protein | |
| g) lettuce | vitamins, minerals and water | |

**Table 4.18**

**1** The Venn diagram shown in figure 4.34 shows the three food types needed for a balanced diet.

Match each food in table 4.18 with its correct number on the Venn diagram. (8)

**2** Table 4.19 shows the mass of each type of food present in 100 g of each of three popular snacks.
a) Which snack contains least fat? (1)
b) Which TWO snacks have the same fibre content and the same carbohydrate content? (1)
c) What is the percentage carbohydrate content of snack A? (1)
d) By how many times is the protein content of snack C greater than that of snack B? (1)
e) What is the whole number ratio of
   (i)   carbohydrate to fibre in snack A?
   (ii)  carbohydrate to protein in snack C? (2)
f) Draw a bar chart of the ingredients that make up 100 g of snack A. (3)

**3** People suffer health problems when they are short of vitamins in their diet. Use the chart in figure 4.35 to answer the following questions.

| food type | mass of food type per 100 g of snack | | |
|---|---|---|---|
| | snack A | snack B | snack C |
| protein | 10 | 4 | 8 |
| carbohydrate | 56 | 60 | 60 |
| fat | 25 | 24 | 22 |
| fibre | 4 | 4 | 4 |
| others | 5 | 8 | 6 |

**Table 4.19**

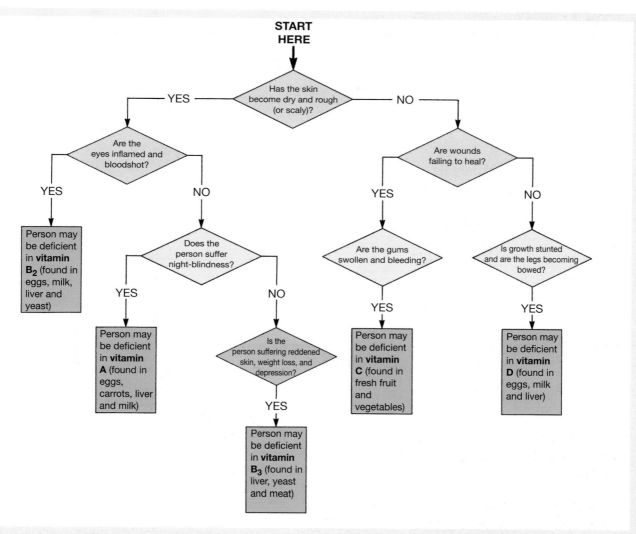

**Figure 4.35**

a) Over 200 hundred years ago, sailors on extended sea voyages often used to suffer soft, bleeding gums. Although the texture of their skin remained normal, any cuts that they got failed to heal.
   (i)   Which vitamin were they short of in their diet?
   (ii)  What food should they have eaten to get a supply of the missing vitamin? (2)

b) If a group of people develop rough, dry skin and have great difficulty seeing in very dim light, which vitamin might be missing from their diet? (1)

c) Rickets is the name given to the deficiency disease suffered by children whose bones are so soft that they bend under the weight of the body.
   (i)   Which vitamin were they short of in their diet?
   (ii)  Which food should be included in their diet to supply the vitamin? (2)

e) According to the chart, which TWO foodstuffs together would provide a rich supply of all five vitamins referred to in the chart? (2)

4 The graphs in figure 4.36 show the relationship between body weight and height for adult males and females.

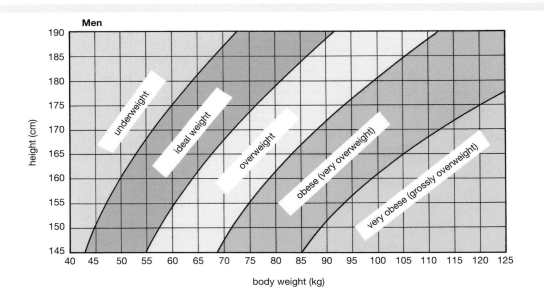

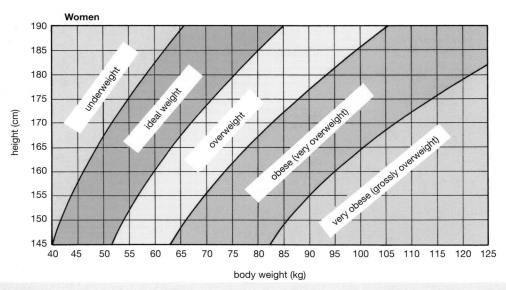

**Figure 4.36**

a) Use these charts to work out the weight rating for each of the five people in table 4.20. (5)

b) Choose the ONE correct answer to each of the following:

(i) A very obese man weighed 100 kg. His height (in cm) could be
**A** 155 **B** 165 **C** 175 **D** 185

(ii) An obese woman's height was 165 cm. Her weight (in kg) could be
**A** 55 **B** 65 **C** 75 **D** 85

(iii) An underweight man weighed 60 kg. His height (in cm) could be
**A** 150 **B** 165 **C** 175 **D** 185

(iv) A woman whose weight was ideal for her height measured 155 cm. Her weight (in kg) could be
**A** 45 **B** 60 **C** 75 **D** 90 (4)

| person | sex | height (cm) | weight (kg) | weight rating |
|--------|--------|-------------|-------------|---------------|
| 1 | female | 180 | 55 | |
| 2 | male | 165 | 110 | |
| 3 | female | 170 | 60 | |
| 4 | male | 175 | 75 | |
| 5 | female | 175 | 75 | |

**Table 4.20**

**5** Read the passage and answer the questions.

## ANOREXIA

Anorexia, the 'slimmers' disease', is a condition mainly affecting young women who begin dieting to lose some weight. The girl soon becomes obsessed with slimming and is so desperate to be thin that she may even starve herself to death.

Often a person suffering from anorexia is very thin and light (e.g. 40 kg or less) and yet when she looks in the mirror, she still thinks that she is too fat (see figure 4.37). This shows that anorexia is a psychiatric condition.

Anorexia is thought to be caused by emotional problems due to factors such as:

- an unhappy home life;
- not wanting to grow up and change from a girl into a woman;
- having parents who are very fat and being determined not to be like them;
- being ignored by potential boyfriends who seem to prefer slimmer girls.

**Figure 4.37**

To cure anorexia, the hidden causes have to be found out first, the recovery is slow and often needs hospital treatment. Good support from family and friends is very important.

a) Describe what the term *anorexia* means. (2)
b) What TWO tell-tale signs would make you suspect that a friend might be suffering from anorexia? (2)
c) Give THREE factors that doctors think may be to blame for anorexia. (3)
d) How do we know that the girl in the picture is suffering from anorexia? (2)

**6** The risk of suffering heart disease depends on several factors and can be calculated by referring to figure 4.38 where each * counts as one risk point.
a) Calculate the level of risk of heart disease for each of the following:
(i)   A woman aged 45 with blood pressure of 120 who eats a diet containing 20% animal fat. Her weight is ideal for her height but she smokes 20 cigarettes a day and takes very little exercise.
(ii)  A 16-year-old with blood pressure of 120 who eats a diet containing 30% animal fat. He is a non-smoker but is 4 kg above his correct weight. He takes a moderate amount of exercise.
(iii) A 61-year-old man who takes very little exercise. He does not smoke but he eats a diet containing 30% animal fat and he is 28 kg above his correct weight. (3)

| age (years) | 10–20 <br> * | 21–30 <br> ** | 31–40 <br> *** | 41–50 <br> **** | 51–60 <br> ***** | 61 and over <br> ****** |
|---|---|---|---|---|---|---|

+

| upper blood pressure (mm Hg) | 100 <br> * | 120 <br> ** | 140 <br> *** | 160 <br> **** | 180 <br> ****** | 200 or more <br> ******* |
|---|---|---|---|---|---|---|

+

| animal fat in diet (%) | 0 <br> * | 10 <br> ** | 20 <br> *** | 30 <br> **** | 40 <br> ***** | 50 <br> ******* |
|---|---|---|---|---|---|---|

+

| variation from ideal weight | 2kg below | 0–2kg above <br> * | 3–9kg above <br> ** | 10–16kg above <br> *** | 17–23kg above <br> ***** | 24kg or more above <br> ****** |
|---|---|---|---|---|---|---|

+

| daily smoking habit | none | cigar/pipe <br> * | 10 cigarettes <br> ** | 20 cigarettes <br> **** | 30 cigarettes <br> ****** | 40 or more cigs. <br> ***** ***** |
|---|---|---|---|---|---|---|

+

| exercise taken | major amounts with workouts <br> * | major amount <br> ** | moderate amount <br> *** | minor amount <br> ***** | very little <br> ****** | none <br> **** **** |
|---|---|---|---|---|---|---|

**key to total ✱ score**
4–11 = below average risk
12–17 = average risk
18–24 = above average risk
above 24 = high risk

**Figure 4.38**

b) A 45-year-old man was found to have a score of 19 risk points. He normally took a moderate amount of exercise, was 4 kg overweight, had a blood pressure of 120 and ate a diet containing 30% animal fat. How many cigarettes did he smoke daily? (1)

c) Name THREE changes to lifestyle that people can make in order to reduce the risk of heart disease. (3)

**7** Figure 4.39 shows how the body responds to increases and decreases in body temperature and returns it to normal.

a) Which part of body senses an increase or decrease in body temperature? (1)

b) When body temperature increases, describe the chain of events that occurs to return the body temperature to normal. (4)

c) When body temperature decreases, describe the chain of events that occurs to return the body temperature to normal. (4)

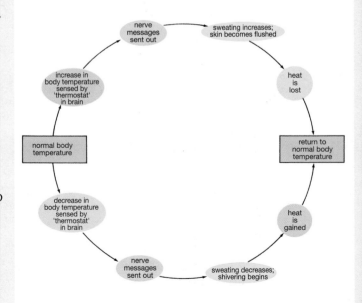

**Figure 4.39**

d) Here are three examples of situations where the system in figure 4.39 failed to work:

1) *The person was suffering hypothermia. She was no longer shivering but had a strong urge to sleep.*

2) *The person was suffering heatstroke as a result of a lengthy spell in a hot, humid environment where he could not cool down by sweating.*

3) *The person was infected with malaria. His body temperature was often as high as 40°C and was accompanied by heavy sweating.*

Match each of these with one of the following first aid procedures that you would give while waiting for expert medical help to arrive:

(i) give the person plenty of cool liquid to drink to replace water lost in sweat;

(ii) wrap the person in extra clothes or a blanket and get them to take a warm drink;

(iii) remove most of the sufferer's clothes and wrap the person in a wet sheet in a cool, airy place. (3)

8 Table 4.21 shows the results of an experiment where a boy used bathroom scales to measure the strength of some of his muscles.

a) Present the data as a bar chart. (3)

b) By how many kg were the finger muscles stronger than the biceps muscles? (1)

| body in action | muscle(s) being used | reading on scale (kg) |
|---|---|---|
| | finger | 42 |
| | biceps | 28 |
| | triceps | 30 |
| | pectoral | 14 |

**Table 4.21**

c) By how many times were the finger muscles stronger than the pectoral muscles? (1)

d) What should now be done to improve the reliability of the results? (1)

**9** In an experiment, a pupil's reaction time was tested repeatedly using a light bulb as a stimulus. As soon as the bulb lit up, the pupil pressed a switch to turn it off. Table 4.22 shows the results.

a) Draw a line graph of the results. Figure 4.40 shows how to add the scale of the vertical (y) axis. (4)

b) (i) In what way was the pupil's reaction time affected by practice?

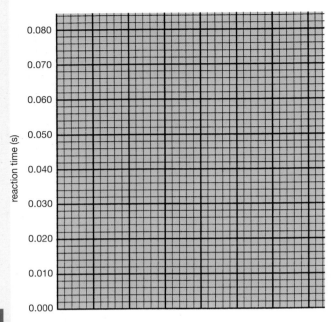

**Figure 4.40**

(ii) By referring to the attempt numbers, state which part of the graph shows this effect. (2)

c) (i) What was the pupil's best reaction time?

(ii) Why is it unlikely that the pupil will be able to improve this time? (2)

**10** Table 4.23 shows the alcohol concentration of the blood of women and men of different weights.

| attempt number | reaction time (seconds) |
|---|---|
| 1 | 0.070 |
| 2 | 0.060 |
| 3 | 0.050 |
| 4 | 0.045 |
| 5 | 0.040 |
| 6 | 0.040 |
| 7 | 0.040 |
| 8 | 0.040 |

**Table 4.22**

| number of units consumed | alcohol concentration of blood soon after consuming drink (mg/cm³ blood) | | | | | |
|---|---|---|---|---|---|---|
| | weight of female (kg) | | | weight of male (kg) | | |
| | 50 | 60 | 75 | 60 | 75 | 90 |
| 1 | 40 | 35 | 28 | 28 | 23 | 19 |
| 2 | 80 | 70 | 56 | 56 | 46 | 38 |
| 3 | 120 | 105 | 84 | 84 | 69 | 57 |
| 4 | 160 | 140 | 112 | 112 | 92 | 76 |
| 5 | 200 | 175 | 140 | 140 | 115 | 95 |
| 8 | 320 | 280 | 224 | 224 | 184 | 152 |
| 10 | 400 | 350 | 280 | 280 | 230 | 190 |

**Table 4.23**

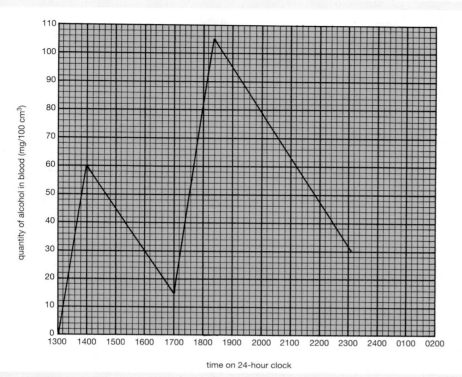

**Figure 4.41**

a) What relationship exists between increase in number of units of alcohol consumed and alcohol concentration in the blood? (1)

b) What effect does increase in body weight have on alcohol concentration in the blood of men when five units have been consumed? (1)

c) Compared with men who weigh 60 kg, what effect does drinking four units of alcohol have on women who also weigh 60 kg? (1)

d) It is illegal to drive when the alcohol concentration of the blood is greater than 80 mg/100 cm³ blood. According to table 4.23, which of the following people should not drive?

  (i)   A 60 kg man who has drunk three units.

  (ii)  A 60 kg woman who has drunk two units.

  (iii) A 90 kg man who has drunk four units.

  (iv) A 75 kg woman who has drunk five units. (4)

**11** The graph in figure 4.41 shows the results of monitoring the blood alcohol level of a man who drank 1.5 pints of beer at lunchtime and 2 pints after work on the same day.

a) The man began drinking the beer at 13.00 hours. By how many mg alcohol/100 cm³ blood did the lunchtime beer make his blood alcohol level rise? (1)

b) (i)  By how much did his blood alcohol level drop between 14.00 and 16.00 hours?

  (ii) By how much did his blood alcohol level drop *per hour* between 14.00 and 16.00? (2)

c) (i)  At what time did he next start drinking alcohol that same day?

  (ii) What was the highest quantity of alcohol found in his bloodstream during all of the time shown in the graph? (2)

d) It is illegal to drive when the blood alcohol level is above 80 mg/100 cm³ blood. What is the earliest time at which

**Figure 4.42**

| distance between hand and stick (mm) | reaction time (seconds) | | |
|---|---|---|---|
| | average for 15 year-old | average for 40 year-old | average for 40 year-old after 4 pints of beer |
| 100 | 0.19 | 0.25 | 0.45 |
| 200 | 0.21 | 0.28 | 0.50 |
| 500 | 0.27 | 0.37 | no result |

**Table 4.24**

the man could legally drive after 18.00 hours? (1)

e) If the trend in the graph continues, what time would it be when the quantity of alcohol in the man's bloodstream reaches zero? (1)

**12** One way of testing how quickly people react to a situation is to time them catching a falling metre stick. An investigation was carried out where the person's hand began at a certain distance away from the metre stick about to be dropped. The person had to reach out and grab the falling stick (see figure 4.42). Table 4.24 shows a typical set of results.

a) Draw a line graph of these results. Figure 4.43 shows how to add the scales to the two axes. (4)

b) (i) Extend each line graph to cut the vertical axis in order to find out what each person's reaction time would have been at zero distance from the metre stick.

(ii) Whose reaction time is shortest? (4)

c) (i) A teenager was found to have a reaction time of 0.22 seconds over a distance of 400 mm. Is this short, long or average for his age?

(ii) A 40-year-old man was found to have a reaction time of 0.34 seconds over a distance of 400 mm. Is this short, long or average for his age? (2)

d) (i) Suggest the reason for the 'no result' entry in the table.

(ii) What effect does four pints of alcoholic drink have on a 40-year-old man's reaction time at the other shorter distances? (2)

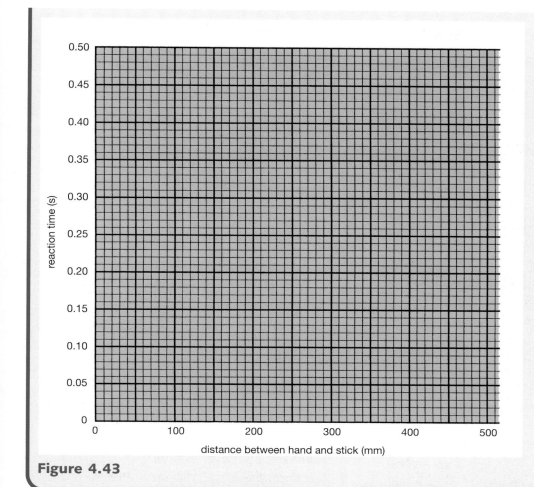

**Figure 4.43**

## What you should know

| | | |
|---|---|---|
| accident | fatigued | oxygen |
| anorexia | fever | practice |
| balance | glass | pregnant |
| blood | growth | reaction |
| calliper | heart | rest |
| control | heatstroke | ruler |
| diabetes | height | sensor |
| digital | hypothermia | strength |
| diseases | liver | temperature |
| dynamometer | long | thermometer |
| elderly | muscles | time |
| energy | nervous | vitamins |

**Table 4.25**  *Word bank for chapter 4*

1  For good health, the body needs a diet that contains a _____ of the three food types. These are fats and carbohydrates to give the body _____, proteins needed for _____ and tissue repair and _____ and minerals to protect the body against deficiency _____.

2  Body fat can be measured using a low-tech instrument called a skin fold _____ or a high-tech instrument called a fat _____. The fat content of a healthy adult's body is about 15–20% for men and 20–25% for women if they are at their ideal body mass. Ideal body mass depends on factors such as _____, age and sex.

3  People who are overweight run an increased risk of suffering _____ disease, kidney failure, arthritis and _____. People who are seriously underweight may be suffering _____ or cancer.

4  The temperature of a healthy human body is around 37°C. Body _____ can be measured using a low-tech instrument called a _____ clinical thermometer or a high-tech instrument called a _____ clinical _____.

5  Body temperatures of 40°C and above can lead to _____ or indicate _____. Body temperatures of 35°C and below indicate _____. Babies and _____ people are particularly at risk from hypothermia.

6  Regular exercise is needed to maintain the size and strength of _____. In the absence of exercise, their size and _____ decrease. The strength of a muscle can be measured using a _____.

7  During exercise, the uptake of _____ by muscles increases. If muscles do not get enough oxygen during exercise, they become _____. This fatigue is relieved by a period of _____.

8  The time taken by the body to respond to a stimulus is called _____ time. It can be measured using a dropped _____ or an electronic reaction timer. Length of reaction time can be reduced by _____. Reaction time is affected by drugs, alcohol and excitement.

9  Reaction time is an indicator of health. If it is _____, this can indicate a problem such as diabetes, a disorder of the _____ system or arterial disease.

10  On being consumed, alcohol is absorbed into the _____ and carried round the body. It causes muscle _____ to become poorer, reaction _____ to become longer and judgements to be less reliable. All of these increase the chance of the person being involved in an _____. Alcohol and drugs consumed by a _____ woman can damage the health of the baby. In the long term, drinking alcohol to excess damages the _____ and brain.

# Unit 2

# Biotechnological Industries

**Biotechnological processes employ the cells of micro-organisms to make products that are so useful to mankind that the quality of people's life is improved.**

## 5 Dairy industries

## Milk

Normal full-fat **milk** (also called whole milk) is a nourishing food and can form an important part of a **balanced diet**. It contains sugar, fats, proteins, vitamins and minerals (see figure 5.1).

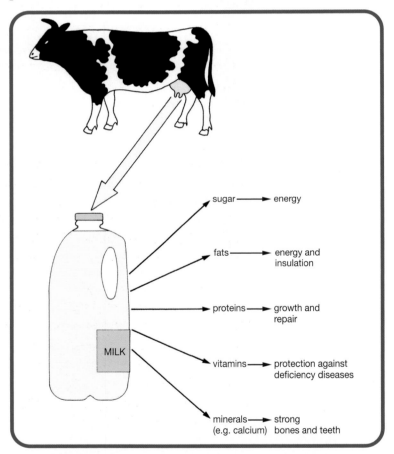

sugar ⟶ energy

fats ⟶ energy and insulation

proteins ⟶ growth and repair

vitamins ⟶ protection against deficiency diseases

minerals ⟶ strong (e.g. calcium) bones and teeth

MILK

**Figure 5.1** *Composition of milk*

## Processing treatments

Different processing treatments are used to produce different types of milk (see figure 5.2).

**Figure 5.2** *Different types of milk*

## Removal of fat

**Skimmed** milk is produced by removing nearly all of the fat from whole milk. **Semi-skimmed** milk is produced by removing about half of the fat from whole milk. These types of milk are popular with people who are trying to control their intake of animal fat and prevent themselves from gaining weight. **Full-cream** milk which is rich in nutrients and energy is recommended for growing children.

## Bacteria

**Bacteria** are tiny, single-celled micro-organisms (microbes) that are widely spread throughout the environment. A few varieties are harmful because they cause diseases if they get into the human body. Most types are completely harmless to the human body and a few are even useful, such as the bacteria used to make yoghurt (see page 108) and cheese (see page 119).

## Heat treatment to kill microbes

A healthy cow normally produces milk that is **sterile** (free from microbes). However if the milk is carelessly handled or stored in unsterilised containers, it can easily become **contaminated** with microbes (e.g. bacteria). Milk is therefore heat-treated before use to destroy harmful microbes. Two methods of treatment are pasteurisation and ultra heat treatment.

### Pasteurisation

During **pasteurisation**, milk is heated to 72°C for 15 seconds (and then quickly cooled down to around 5°C). This process destroys disease-causing microbes. However it does not kill all of the bacteria that make milk turn sour. So an unopened container of pasteurised milk will go sour eventually but takes about five days longer than a sample of unpasteurised milk. (Pasteurised milk must be refrigerated and used within a few days.)

### UHT (ultra heat treatment)

This is a process by which milk is heated to a high temperature such as 138°C for 2–5 seconds. This heat process destroys the disease-causing microbes *and* the bacteria that make milk turn sour. **UHT** treatment is therefore used to **preserve** milk and prolong its shelf-life. An unopened container of UHT milk will last for up to six months without there being any chance of it going off. Once opened, it must be refrigerated and used within a few days.

## Evaporation

Evaporated milk is produced by heating milk to remove more than half of its water content. This makes the milk more concentrated. It is a further method of **preserving** milk because bacteria are unable to survive in it.

## Taste

In every case the processing treatment provides us with benefits. However the taste of the milk, compared with milk fresh from a cow, is different. Pasteurisation only affects the taste slightly but UHT makes a significant difference to it.

## Resazurin test

**Resazurin** dye is a chemical that changes colour in response to the number of bacteria present in the liquid to which it has been added. Resazurin can be used to indicate the bacterial content of milk as shown in table 5.1. The faster the colour change from **blue-purple** to white, the greater the number of bacteria present in the milk.

| colour of milk sample containing resazurin dye after 15 min at 37°C | bacterial content of milk | drinking quality of milk |
| --- | --- | --- |
| blue-purple | very low | good |
| mauve | low | satisfactory |
| pink | medium | poor |
| white | high | unsatisfactory |

**Table 5.1**   *Resazurin dye test*

### Using resazurin to test milk for bacteria

In this experiment, resazurin dye is added to equal volumes of fresh milk, 5-day-old milk and 10-day-old milk. The starting colour of each test tube's contents is noted. The test tubes are then placed in a water bath at 37°C as shown in figure 5.3. The colour of each tube is noted at 5-minute intervals for 15 minutes.

Table 5.2 gives a typical set of results. From this experiment it is concluded that fresh milk has the lowest bacterial content and 10-day-old milk has the highest bacterial content. It is also concluded that the **drinking quality** of fresh milk is *good*, 5-day-old milk *poor* and 10-day-old milk *unsatisfactory*.

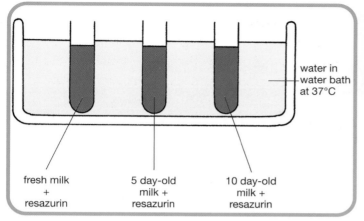

fresh milk + resazurin     5 day-old milk + resazurin     10 day-old milk + resazurin

**Figure 5.3**   *Testing milk for bacteria*

| time that sample has been in water bath (min) | fresh milk | 5 day-old milk | 10 day-old milk |
|---|---|---|---|
| 0 | | | |
| 5 | | | |
| 10 | | | |
| 15 | | | |

**Table 5.2**   *Results of resazurin dye test*

## Activity

## Resazurin test (assessment)

You need
- four labels
- four boiling tubes
- stand for boiling tubes
- 10 cm³ sample of each of the following (stored in fridge at 4°C):
  - fresh milk
  - 3-day-old milk
  - 6-day-old milk
  - 12-day-old milk
- 25 cm³ resazurin dye
- four rubber stoppers
- access to water bath at 37°C
- stopclock

What to do
1. Make a copy of table 5.3 before starting the experiment.
2. Label the four boiling tubes: *fresh*, *3-day-old*, *6-day-old* and *12-day-old*.
3. Add each milk sample to the appropriate boiling tube.
4. Add 5 cm³ resazurin dye to each milk sample.
5. Insert the rubber stoppers and turn each tube upside down twice to mix the contents.
6. Note the starting colour of each tube and enter it in your table.
7. Place the four tubes in the water bath and start the clock.
8. Once the experiment has been running for 5 minutes, enter the colour of each tube in your table.

9  Repeat this procedure at 10 minutes and at 15 minutes.

10  Draw a conclusion from your results at 15 minutes about the bacterial content and drinking quality of each of the four milk samples.

11  Ask the teacher to assess your work by checking your four tubes at 15 minutes.

| time that sample has been in water bath (min) | colour of resazurin dye in milk | | | |
|---|---|---|---|---|
| | fresh milk | 3-day-old milk | 6-day-old milk | 12-day-old milk |
| 0 | | | | |
| 5 | | | | |
| 10 | | | | |
| 15 | | | | |

**Table 5.3** *Resazurin dye test results*

# Yoghurt

**Yoghurt** (see figure 5.4) is a food produced from milk by the action of *yoghurt-forming* bacteria. The flow diagram in figure 5.5 shows a simplified version of a procedure carried out to produce yoghurt on a commercial scale.

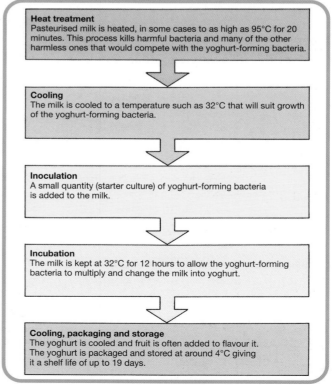

**Heat treatment**
Pasteurised milk is heated, in some cases to as high as 95°C for 20 minutes. This process kills harmful bacteria and many of the other harmless ones that would compete with the yoghurt-forming bacteria.

**Cooling**
The milk is cooled to a temperature such as 32°C that will suit growth of the yoghurt-forming bacteria.

**Inoculation**
A small quantity (starter culture) of yoghurt-forming bacteria is added to the milk.

**Incubation**
The milk is kept at 32°C for 12 hours to allow the yoghurt-forming bacteria to multiply and change the milk into yoghurt.

**Cooling, packaging and storage**
The yoghurt is cooled and fruit is often added to flavour it. The yoghurt is packaged and stored at around 4°C giving it a shelf life of up to 19 days.

**Figure 5.4**  *Yoghurt*

**Figure 5.5**  *Yoghurt production*

During the incubation stage of this process, the sugar in the milk is used by the yoghurt-forming bacteria and converted to an acid called **lactic acid**. Lactic acid *clots* and *thickens* the milk and gives yoghurt its tangy, 'sour' flavour. As the acid builds up, the pH of the yoghurt drops to about 5.5.

## Preservative

Since other bacteria that make food go bad cannot grow in acidic conditions, lactic acid is described as a **natural preservative**. Making yoghurt is a further method of preserving milk.

Many types of yoghurt (e.g. these labelled 'natural' yoghurt and 'BIO' yoghurt on the carton) still contain *live* yoghurt-forming bacteria. These are harmless if consumed. However, if the yoghurt is not stored in a cool place, these bacteria continue to multiply and make more lactic acid. Soon the yoghurt becomes too sour to eat.

Natural yoghurt containing live yoghurt-forming bacteria can be used as a source of a **starter culture** of bacteria to make more yoghurt (see page 111).

## Activity

### Examining yoghurt-forming bacteria

**You need**
- spatula
- carton of natural yoghurt
- test tube
- two microscope slides
- dropping bottle of nigrosin stain
- microscope and light source

**What to do (also see figure 5.6)**
1 Using the spatula, add a drop of natural yoghurt to a little water in a test tube and stir.
2 Add a drop of the diluted yoghurt to a glass slide.
3 Add a drop of nigrosin stain.
4 Using the spatula, mix the yoghurt and stain.
5 Using the second glass slide, drag the mixture along the slide leaving a thin layer.
6 Allow the specimen to dry for 5 minutes.
7 Set up the microscope.

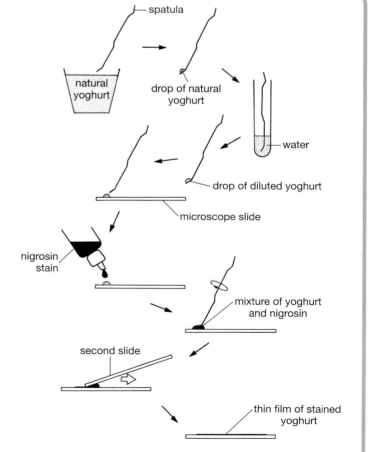

**Figure 5.6** *Preparing a slide of yoghurt bacteria*

8 Examine the specimen under the highest magnification and look for bacteria.

9 Compare your findings with the bacteria shown in figure 5.7.

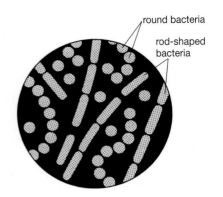

round bacteria

rod-shaped bacteria

**Figure 5.7** *Yoghurt-forming bacteria*

# Aseptic conditions

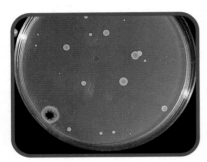

**Figure 5.8** *Air-borne bacteria and fungi*

Air contains various types of microbe. Many of these are present in dust as tiny structures called **spores**. These air-borne spores are constantly landing on all surfaces. Figure 5.8 shows a Petri dish of jelly-like food called **nutrient agar** on which spores from the air have landed and started to grow.

During laboratory work with micro-organisms, certain precautions are taken to try to create **sterile** (**aseptic**) conditions in which to carry out the experiment. This is done for two reasons:

- to stop unwanted microbes getting into the experiment and spoiling it;
- to stop the microbe used in the experiment from escaping into the environment.

## Work surfaces and microbiologists

All work surfaces in a microbiology laboratory should be made of a hard, non-absorbent material such as glass or plastic. Hands should be washed, any cuts should be covered with waterproof dressings and lab coats should be worn to prevent contamination. There should be no eating or drinking in the laboratory.

## Sterile equipment

A few microbes can even survive boiling water. Equipment to be used in a microbiology experiment should therefore be heated under pressure to 121°C for 20 minutes in a special container called an

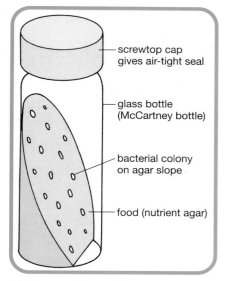

**Figure 5.9** *Culture bottle of a microbe*

autoclave (see page 132). This treatment makes sure that all microbes are dead and therefore that the equipment is sterile at the start of the experiment.

## Pure culture

A pure culture of a microbe (e.g. the yoghurt-forming bacterium *Lactobacillus*) is often grown inside a small glass bottle (see figure 5.9) on nutrient agar.

## Inoculating loop

An instrument is needed to transfer some of the microbes from a pure culture bottle to another site (e.g. a sterilised test tube containing pasteurised milk). This instrument is called an **inoculating loop** (see figure 5.10).

### Sterilising an inoculating loop

The wire part of the loop is heated to *red heat* in a Bunsen flame to kill all the microbes. This is done by heating the shaft near the handle first and then drawing the rest of the wire through the flame until all parts have been red hot (see figure 5.11). After this **flaming** process, the loop is allowed to cool for a few seconds before use (but it must not be laid down on a non-sterile surface.

**Figure 5.10** *Wire inoculating loop*

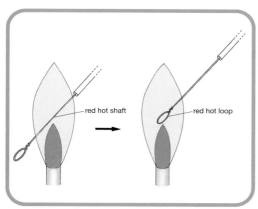

**Figure 5.11** *Sterilising an inoculating loop by flaming*

### Sub-culturing a micro-organism

- The work is done close to a Bunsen burner so that the draught from the flame will sweep stray microbes in the air upwards and away from the experiment.
- The cap on the culture bottle is loosened but not yet taken off.
- The bottle is held by one hand and the cap removed and held by the little finger of the other hand (see figure 5.12).
- The neck of the bottle is flamed in the Bunsen to kill any microbes present there.
- A sterilised inoculating loop is then used to gently scrape a small sample of microbes from the surface of the nutrient agar in the culture bottle.
- The neck of the culture bottle is flamed again and its cap replaced.
- The wire loop carrying the microbes is used to inoculate the new container of growing medium.
- The wire loop is reflamed to kill microbes.

Figure 5.12 shows this procedure being used to inoculate pasteurised

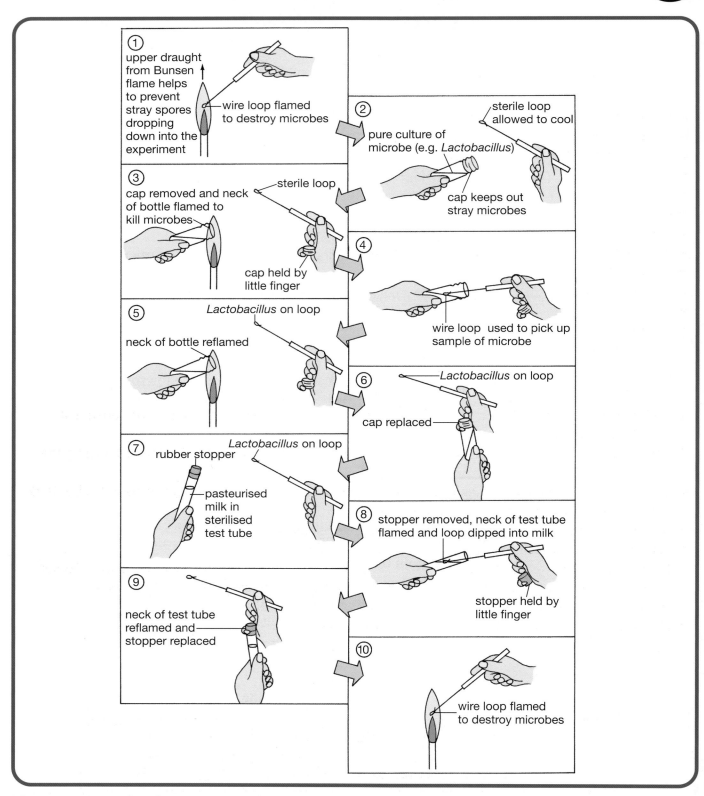

**Figure 5.12** *Sub-culturing a micro-organism*

milk with yoghurt-forming bacteria (*Lactobacillus*) under aseptic conditions. If the test tube of pasteurised milk is kept at 32°C for 12 hours, the milk is found to thicken up into yoghurt.

## Activity

# Producing yoghurt

*You need*

- six labels
- six test tubes
- test tube stand
- pasteurised milk
- UHT milk
- dried milk 'dissolved' in water
- two metal spatulas
- natural yoghurt (rich in live yoghurt-forming bacteria)
- boiled and cooled natural yoghurt
- six rubber stoppers
- sellotape
- scissors
- access to water bath at 32°C

*What to do (see also figure 5.13)*

1 Label the six test tubes A, B, C, D, E and F and add your initials.
2 Half-fill tubes A and B with pasteurised milk, C and D with UHT milk and E and F with dried milk 'dissolved' in water.
3 Using a spatula, add a large scoop of natural yoghurt to test tubes A, C and E without touching the milk with the spatula.
4 Using the second spatula, add an equal scoop of boiled and cooled natural yoghurt to test tubes B, D and F without touching the milk with the spatula.
5 Seal each tube with a rubber stopper and sellotape and then shake it thoroughly.
6 Place the six test tubes in the water bath at 32°C for 12 hours.
7 Next lesson, examine your tubes and find out which milk samples have turned into yoghurt. Do *not* open the tubes and do *not*, under any circumstances, taste the contents of any of the tubes.
8 Draw a conclusion about which types of milk are suitable for making yoghurt.
9 Return your sealed tubes to your teacher for safe disposal.

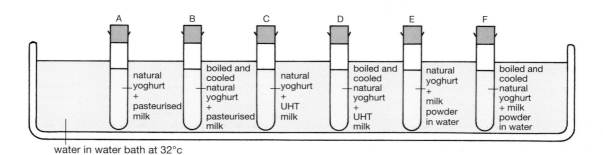

**Figure 5.13** *Making yoghurt*

Activity

## Investigating the effect of a starter yoghurt culture on the pH of milk

### You need

- two labels
- two test tubes
- test tube stand
- UHT milk
- two metal spatulas
- natural yoghurt (rich in yoghurt-forming bacteria)
- boiled and cooled natural yoghurt
- dropping bottle of pH indicator solution
- two rubber stoppers
- pH reference scale
- *optional if available*: pH sensor attached to a computer (e.g. Pasco equipment)
- access to incubator at 32°C

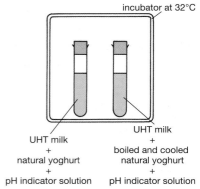

incubator at 32°C

UHT milk
+
natural yoghurt
+
pH indicator solution

UHT milk
+
boiled and cooled
natural yoghurt
+
pH indicator solution

**Figure 5.14**   *Investigating the effect of yoghurt on the pH of milk*

| when pH was recorded | A | B |
|---|---|---|
| at start | | |
| after 2 days at 32°C | | |

**Table 5.4**   *Milk pH results*

### What to do (also see figure 5.14)

1. Prepare a copy of table 5.4.
2. Label the test tubes A and B and add your initials.
3. Half-fill each tube with UHT milk.
4. Using a spatula, add a large scoop of natural yoghurt to tube A.
5. Using the second spatula, add an equal scoop of boiled and cooled natural yoghurt to tube B. (*Note*: Tube B is called the **control**. It is a copy of the experiment where all factors are kept the same except the one being investigated. When the tubes are compared at the end of the experiment, any difference between them must be due to that one factor.)
6. Add 20 drops of pH indicator solution to each tube.
7. Seal each tube with a rubber stopper and shake it thoroughly.
8. Using the pH scale, record the pH of the contents of each tube in your table. (If available, also use a pH sensor attached to a computer to read each pH.)
9. Place the two test tubes in an incubator at 32°C.
10. After two days, repeat step 8.
11. Draw a conclusion from your results.
12. Return your stoppered tubes to the teacher for safe disposal.

## Testing your knowledge

**1** Name FOUR classes of food found in whole milk. (4)

**2** In what way must whole milk be treated to change it into skimmed milk? (1)

**3** a) Match boxes (i), (ii) and (iii) in figure 5.15 with boxes A, B and C. (3)
   b) Which of these forms of heat treatment kills disease-causing microbes but does not kill all of the bacteria that make milk turn sour? (1)

| form of heat treatment | description of process |
|---|---|
| (i) ultra heat treatment (UHT) | A milk heated to remove more than half of its water content |
| (ii) evaporation | B milk heated to 72°C for 15 seconds |
| (iii) pasteurisation | C milk heated to 138°C for 4 seconds |

**Figure 5.15**

**4** a) Briefly describe how you would carry out the resazurin test on a milk sample. (2)
   b) If the sample turned pink, what would this result indicate about:
   (i)  the bacterial content of the milk?
   (ii) the drinking quality of the milk? (2)
   c) If milk has a very low bacterial content and good drinking quality, what colour would result when it was tested with resazurin? (1)

**5** a) (i)  When yoghurt-forming bacteria are added to milk, which food in the milk do they use up?
   (ii) Which chemical substance do they change this food into?
   (iii) What effect does this chemical have on the milk? (3)
   b) Explain how lactic acid acts as a natural preservative. (1)
   c) Why does natural yoghurt, if left out of the fridge for a few days, become too sour to eat? (1)

## Cheese

Cheese is a *solid* food. It is made from milk, a *liquid* food. For this to be possible, the milk must first be made to **curdle**. This means changing it into solid white lumps called **curds** and a clear liquid called **whey** (see figure 5.16).

**Figure 5.16** *Fresh and curdled milk*

Activity

### Investigating the effect of adding rennet to milk

You need
- two labels
- two boiling tubes
- measuring cylinder (10 cm³ plastic)
- pasteurised whole milk
- dropping bottle of rennet *or* fungal rennet (fromase) *or* GM (maxiren) rennet
- dropping bottle of water
- two rubber stoppers
- access to water bath at 37°C

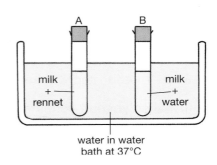

**Figure 5.17** *Investigating the effect of adding rennet to milk*

What to do (also see figure 5.17)

1 Label the two boiling tubes A and B and add your initials.
2 Using the measuring cylinder, add 10 cm³ of milk to each tube.
3 Add 25 drops of rennet to tube A and 25 drops of water to tube B.
4 Fit stoppers to both tubes and invert them three times to mix the contents.
5 Stand the two tubes in the water bath at 37°C.
6 After 20 minutes, inspect the tubes by inverting them and then comparing results with other students who have used different forms of rennet.
7 Write a sentence to say what has happened.
8 Draw a conclusion about the effect of rennet on milk.
9 Keep your stoppered boiling tubes set up and leave them at room temperature until your next Biology lesson.

# Three food tests

### Testing for starch

A small sample of the food material to be tested is placed in a test tube. A few drops of **iodine solution** are added to the food. If a **blue–black** colour is produced, then starch is present in the food.

### Testing for protein

A small sample of the food material to be tested is placed in a test tube. A few drops of **Biuret reagent** are added to the food. If a **lilac** colour is produced, then protein is present in the food.

### Testing for simple sugar

A small sample of the food to be tested is placed in a test tube. A few drops of blue **Benedict's solution** are added to the food. The test tube is placed in a water bath at 90°C for a few minutes. If an **orange** colour is produced, then simple sugar is present in the food.

The three tests are summarised in figure 5.18.

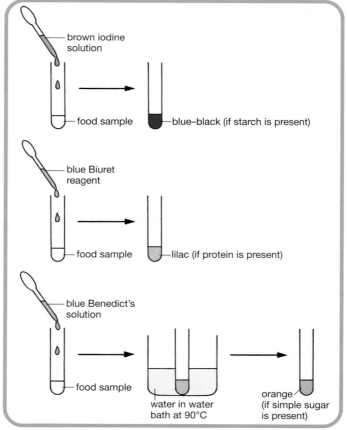

**Figure 5.18** *Three food tests*

## Activity

# Carrying out food tests on curds and whey

### You need

- tea strainer
- two beakers (250 cm³ plastic)
- stoppered boiling tube A from previous activity
- six labels
- six test tubes
- metal spatula
- dropping bottle of iodine solution
- dropping bottle of Biuret reagent
- dropping bottle of Benedict's solution
- two test tube holders
- access to water bath at 90°C

### What to do (also see figure 5.19)

1 Make a copy of table 5.5.
2 Balance the strainer on top of one of the beakers.
3 Carefully pour the contents of tube A into the strainer so that the whey passes through and the curds are caught.
4 Tip the curds into the second beaker.
5 Label the six test tubes 1–6.
6 Using the spatula, add an equal scoop of curds to test tubes 1, 2 and 3.
7 One-quarter fill test tubes 4, 5 and 6 with whey.
8 Test the curds and whey for starch by adding three drops of iodine solution to test tubes 1 and 4. (Positive result = *blue–black*.)

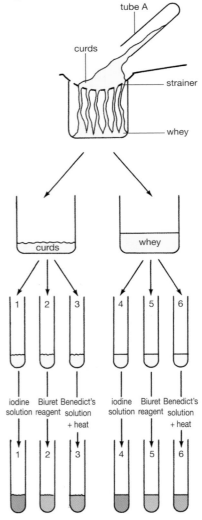

**Figure 5.19** *Carrying out food tests on curds and whey*

| test tube | part of milk present | food being tested for | final colour | food present or absent |
|-----------|----------------------|-----------------------|--------------|------------------------|
| 1 | curds | starch | | |
| 2 | curds | protein | | |
| 3 | curds | simple sugar | | |
| 4 | whey | starch | | |
| 5 | whey | protein | | |
| 6 | whey | simple sugar | | |

**Table 5.5** *Food test results*

9  Test the curds and whey for protein by adding six drops of Biuret reagent to test tubes 2 and 5. (Positive result = *lilac*.)

10  Test the curds and whey for simple sugar by adding six drops of Benedict's solution to tubes 3 and 6. Hold both tubes in the water bath for a few minutes using test tube holders. (Positive result = *orange*.)

11  Complete the table to show your results.

12  Draw a conclusion from your experiment about classes of food present in curds and whey.

## Activity

# Investigating the effect of pH on the activity of rennet

*You need*

- three labels
- three boiling tubes
- stand for boiling tubes
- measuring cylinder (10 cm³ plastic)
- pasteurised whole milk
- dropping bottle of pH indicator solution
- dropping bottle of 1M hydrochloric acid
- dropping bottle of water
- dropping bottle of 1M sodium hydroxide (alkali)
- dropping bottle of rennet
- three rubber stoppers
- pH reference scale
- *optional if available*: pH sensor attached to a computer (e.g. Pasco equipment)
- access to water bath at 37°C
- stopclock

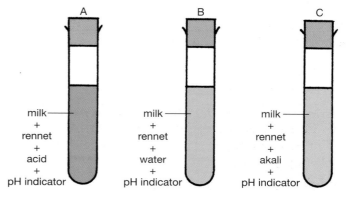

**Figure 5.20**  *Investigating the effect of pH on rennet activity*

*What to do* (also see figure 5.20)

1  Prepare a copy of results table 5.6.

2  Label the boiling tubes A, B and C and add your initials.

3  Using the measuring cylinder, add 10 cm³ of milk to each tube.

4  Add 20 drops of pH indicator solution to each tube.

5  Add 10 drops of acid to tube A.

6  Add 10 drops of water to tube B.

7  Add 10 drops of alkali to tube C.

8  Add 25 drops of rennet to each tube.

9  Put a rubber stopper into each tube and invert it to mix the contents.

10  Using the pH reference scale, record in your table the pH of the contents of each tube. (If available also use a pH sensor attached to a computer to read each pH.)

11  Place the three tubes in the water bath at 37°C and start the stopclock.

**12** Examine the contents of each tube at 2-minute intervals by tilting them gently to see if the contents have clotted.

**13** Complete the results table as you go along.

**14** Draw a conclusion about the effect of pH on the activity of rennet on milk.

| time (min) | tube | | |
| --- | --- | --- | --- |
| | A (pH = ) | B (pH = ) | C (pH = ) |
| 0 | ✕ | ✕ | ✕ |
| 2 | | | |
| 4 | | | |
| 6 | | | |
| 8 | | | |
| 10 | | | |

key    ✓✓ = complete clotting
        ✓ = some clotting
        ✕ = no clotting

**Table 5.6** *Results for effect of pH on rennet activity*

## Rennet

**Rennet** is a chemical substance which makes milk **curdle**. It does this by clotting the protein in milk into **curds.** It is therefore used in the production of cheese (see page 120).

### Sources of rennet

#### Calf rennet
Rennet is made naturally by the stomachs of young mammals that are still feeding from their mothers. The first source of rennet used in cheese-making was extracted from the **stomach lining** of calves.

#### Fungal rennet
Some fungi are found to make rennet naturally. These fungi are grown on a large scale in a **fermenter** (see page 193) and the rennet is extracted and purified ready for use in the cheese-making industry.

#### Rennet from genetically engineered yeast
Yeast is a unicellular fungus normally used in baking and brewing (see chapter 6). In recent times, scientists have produced a **genetically engineered** strain of yeast that can make rennet.

They have done this by taking the gene that tells calf stomach cells how to make the rennet chemical and putting the gene into the yeast cells. These **genetically modified** yeast cells are grown on a massive scale in a fermenter. They produce rennet that is chemically identical to calf rennet and ideal for use in the cheese-making industry.

## Advantages and disadvantages

Advantages and disadvantages of the three types of rennet are summarised in table 5.7.

| | advantages | disadvantages |
|---|---|---|
| **calf rennet** | a) It is the original natural form of the chemical substance.<br>b) It does the same job in the production of cheese as it did in the stomachs of calves. | It is a substance that can only be obtained by slaughtering animals and this is unacceptable to many people. |
| **fungal rennet** | a) It is the cheapest form of rennet.<br>b) It does not involve the slaughter of animals and it therefore allows vegetarian cheese to be made.<br>c) It is a natural substance found in the wild. | It continues to break down protein when the cheese is made and this can cause a poor flavour to develop. |
| **rennet from genetically engineered yeast** | a) It does the same job in the production of cheese as calf rennet.<br>b) It does not involve the slaughter of animals and it therefore allows vegetarian cheese to be made. | It is formed by a genetically modified organism and it is therefore unacceptable to some people who feel that it is unnatural. |

**Table 5.7**  *Advantages and disadvantages of the three types of rennet*

# Cheese-making bacteria

In addition to rennet, the production of cheese depends on the action of special strains of **bacteria** (see figure 5.21). Starter cultures of these bacteria are added to the milk. They multiply rapidly and change milk sugar into **lactic acid**. This acid helps to **clot** the milk protein into curds. The bacteria also affect the final flavour of the cheese. The manufacture of cheese is shown in figures 5.22 and 5.23.

**Figure 5.21**  *Cheese-making bacteria*

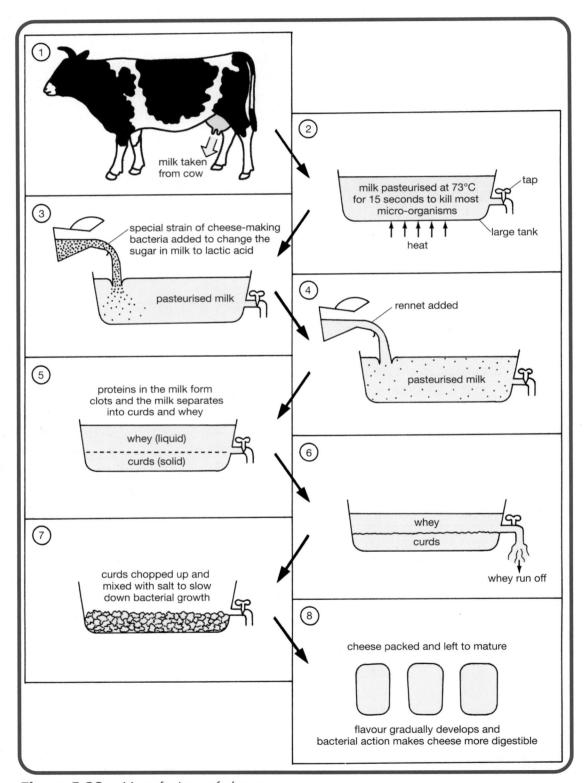

**Figure 5.22** *Manufacture of cheese*

**Figure 5.23**  *Cheese-making*

# Environmental impact

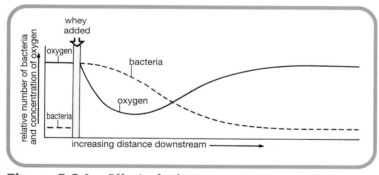

**Figure 5.24**  *Effect of whey on a river*

## Potential effect of whey on a river

For many years **whey** produced by the cheese-making industry was regarded as a *waste*. It was often poured into the nearest river. But whey is a foodstuff rich in **sugar** and **protein**. Once in a river, whey acts as a source of food for bacteria already present in low numbers in the river water.

These bacteria increase rapidly in number and use up most of the **oxygen** dissolved in the river water (see figure 5.24). This leads to a dramatic fall in the numbers and types of other living organisms such as fish that depend on this oxygen. Many miles downstream, in the absence of whey, the concentration of oxygen in the water gradually increases until eventually the original level is restored.

## Upgrading whey

Whey is no longer regarded as a useless waste. Instead of being thrown away it can be upgraded. **Upgrading** means changing a waste material into a useful product. This can be done by using the sugar and protein in whey as a food source for growing vast populations of **yeast** cells.

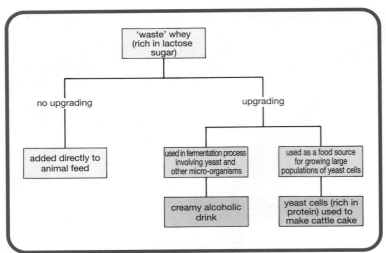

**Figure 5.25**  *Using 'waste' whey*

Some strains of yeast can be grown on whey in a fermenter to produce a creamy **alcoholic drink**. Other strains of yeast fed on whey produce **high-grade protein** that can be used in animal feed (see figure 5.25).

### Direct use of whey
Whey is often sold in liquid form to farmers for use in pig food. Dehydrated whey as whey powder is used as an ingredient in many foods that humans eat such as some types of biscuit.

## Testing your knowledge

**1** a) What does rennet do to milk?  (1)
   b) Under which pH conditions (acidic, neutral or alkaline) does rennet
   (i)  work best?
   (ii) hardly work at all?  (2)
   c) Give TWO differences in appearance between curds and whey.  (2)

**2** Copy and complete table 5.8 which refers to three food tests.  (7)

| food being tested for | reagent used for test | original colour of testing reagent | heat or no heat | colour resulting when food is present (positive result) |
|---|---|---|---|---|
| starch | iodine solution |  | no heat |  |
| protein |  | blue |  | lilac |
| simple sugar |  |  | heat |  |

**Table 5.8**

**3** a) Rennet can be obtained from calves. Give ONE disadvantage of this source.  (1)
   b) Rennet is made naturally by some fungi. Give TWO advantages of this source.  (2)
   c) (i)  State a third source of rennet.
      (ii) Give ONE advantage and ONE disadvantage of this third source.  (3)
                                                                            (4)

**4** If whey is poured into a river, what effect does this have on each of the following?
   a) The number of bacteria in the river water. Explain why.
   b) The concentration of oxygen in the water. Explain why.
   c) The number of fish in the river. Explain why.  (6)

**5** Copy the flow diagram in figure 5.26 and then complete the four blank boxes using the following answers:

(1) curds cut up and mixed with salt

(2) milk pasteurised

(3) whey upgraded to animal feed

(4) rennet added

(4)

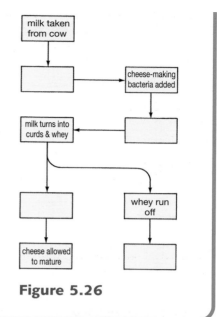

**Figure 5.26**

## Applying your knowledge

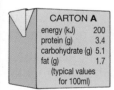 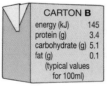 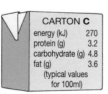

CARTON **A**
energy (kJ)  200
protein (g)  3.4
carbohydrate (g) 5.1
fat (g)  1.7
(typical values for 100ml)

CARTON **B**
energy (kJ)  145
protein (g)  3.4
carbohydrate (g) 5.1
fat (g)  0.1
(typical values for 100ml)

CARTON **C**
energy (kJ)  270
protein (g)  3.2
carbohydrate (g) 4.8
fat (g)  3.6
(typical values for 100ml)

**Figure 5.27**

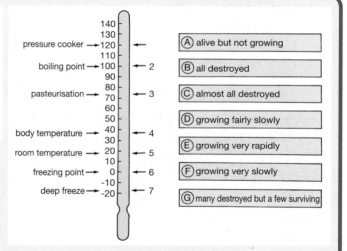

**Figure 5.28**

**1** Figure 5.27 shows three types of milk A, B and C.

a) Match the cartons with the terms *whole milk*, *semi-skimmed milk* and *skimmed milk*. (1)

b) Why is the energy content of milk type C much higher than either of the other two types of milk? (1)

c) (i) In milk type A what is the whole number ratio of carbohydrate to fat?

(ii) In type B what is the whole number ratio of carbohydrate to fat? (2)

**2** In figure 5.28 each pair of arrows points to a particular temperature on the thermometer. Match arrows 1–7 with boxes A–G to show the state of the bacteria in a sample of fresh cow's milk at each of the temperatures. (7)

| milk | fat content | heat treatment given to milk |
|------|-------------|------------------------------|
| UHT skimmed | almost zero | 140°C for 5 seconds |
| UHT semi-skimmed | half | 140°C for 5 seconds |
| UHT whole | full | 140°C for 5 seconds |
| pasteurised skimmed | almost zero | 73°C for 15 seconds |
| pasteurised semi-skimmed | half | 73°C for 15 seconds |
| pasteurised whole | full | 73°C for 15 seconds |

**Table 5.9**

**3** Table 5.9 contains information about six types of milk. Use the information to complete the key of paired statements by supplying the answers to the blank spaces labelled W, X, Y and Z. (4)

**1)** milk heated to 73°C for 15 seconds
.................................................. go to 2)
milk heated to 140°C for 15 seconds
.................................................. go to 4)

**2)** milk with half- or full-fat content
.................................................. go to 3)
milk with almost no fat
........................**pasteurised skimmed milk**

**3)** milk with half of its fat content
.......................... **W**_____
milk with full-fat content
.......................... **X**_____

**4)** milk with half- or full-fat content
.............................................go to 5)
milk with almost no fat
.......................... **Y**_____

**5)** milk with half of its fat content
.......................... **UHT semi-skimmed milk**
**Z**_____
.................................... **UHT whole milk**

**4** Read the following passage and answer the questions on page 125.

## GOATS' MILK

For most people, cows' milk is an excellent source of food. However some people find it difficult to digest; others are allergic to its protein molecules and may suffer asthma, diarrhoea, migraine or other problems when they drink it. Many of these people find that goats' milk provides them with an acceptable substitute. Its composition is very similar to cows' milk (see figure 5.29) yet it is more easily digested and it does not normally produce allergic reactions.

Under natural breeding conditions, female goats give birth in late spring and therefore their milk is plentiful during the summer months. Suppliers often freeze extra goats' milk during the summer to overcome winter shortages. However this practice is becoming less important as farmers can now control the goats' breeding cycle. As a result, at least some of their goats are producing milk at all times of the year. Fresh pasteurised goats' milk is now available all the year round but at present it is about twice as expensive as cows' milk. Goats' milk can also be made into cheese, yoghurt and ice cream.

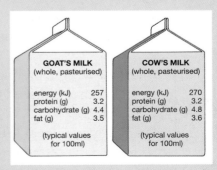

**Figure 5.29**

a) (i)  Which of the two types of milk shown in figure 5.29 has the higher energy content?

   (ii)  With reference to the information in the diagram, give TWO reasons why this is the case. (3)

b) How many grams of protein would be present in 1 litre (1000 cm³) of goats' milk? (1)

c) (i)  Give TWO reactions that could be shown by a person allergic to cows' milk.

   (ii)  Why might the person not quickly realise that these reactions were caused by an allergy to cows' milk? (3)

d) (i)  Many years ago, fresh goats' milk was normally only available during the summer months. Why?

   (ii)  State ONE practice carried out by farmers that ensures that fresh goats' milk is now available all the year round. (2)

e) Give ONE possible advantage and ONE disadvantage of buying goats' milk in place of cows' milk. (2)

**5** Table 5.10 shows the results of testing samples of whole cows' milk from four farms using resazurin dye.

a) (i)  Which farm's milk has the lowest bacterial count?

   (ii)  Explain your answer. (2)

b) (i)  Which farm's milk is unfit for human consumption?

   (ii)  Explain your answer. (2)

c) The farmer whose milk failed the test replied as shown in figure 5.30. What should be done next to see if he is right? (1)

d) Give ONE precaution that scientists should always apply to their equipment when doing the resazurin test. (1)

**6** Natural yoghurt contains live bacteria that change sterile milk into yoghurt. A boy was asked to set up an experiment to find out the best temperature for making yoghurt. He rinsed three glass beakers with warm soapy water and dried them on paper towels. He then used them to set up the experiment shown in figure 5.31.

**Figure 5.30**

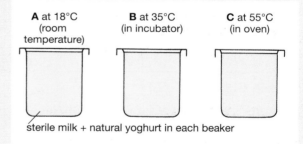

**A** at 18°C (room temperature)   **B** at 35°C (in incubator)   **C** at 55°C (in oven)

sterile milk + natural yoghurt in each beaker

**Figure 5.31**

| time (min) | colour of milk sample containing resazurin dye | | | |
|---|---|---|---|---|
| | farm P | farm Q | farm R | farm S |
| 0 | blue–purple | blue–purple | blue–purple | blue–purple |
| 5 | mauve | blue–purple | blue–purple | blue–purple |
| 10 | pink | blue–purple | mauve | blue–purple |
| 15 | pink | blue–purple | mauve | mauve |
| 20 | white | blue–purple | pink | mauve |

**Table 5.10**

|   | texture of 'yoghurt' | smell of 'yoghurt' |
|---|---|---|
| A | watery | faintly unpleasant |
| B | creamy | unpleasant |
| C | watery | pleasant |

**Table 5.11**

The results after six hours are shown in table 5.11.

a) What was the important error in the boy's procedure that probably led to the unpleasant smell developing? (1)

b) The error was corrected in a second attempt and the results were as shown in table 5.12. The boy concluded that 35°C is the best temperature for making yoghurt. Suggest how his experiment could be improved to find out a more accurate estimate of the very best temperature. (1)

|   | texture of 'yoghurt' | smell of 'yoghurt' |
|---|---|---|
| A | watery | pleasant |
| B | creamy | pleasant |
| C | watery | pleasant |

**Table 5.12**

**7** Figure 5.32 shows a set of experiments to find out the best conditions for making milk clot using rennet.

a) Give the TWO experiments that should be compared to find out:
  (i) which type of rennet clots milk faster at 20°C and pH 6;
  (ii) which type of rennet clots milk faster at 37°C and pH 9. (2)

b) Give the TWO experiments that should be compared to find out:
  (i) which temperature makes calf rennet clot milk faster at pH 9;
  (ii) which temperature makes fungal rennet clot milk faster at pH 6. (2)

c) Give the TWO experiments that should be compared to find out:

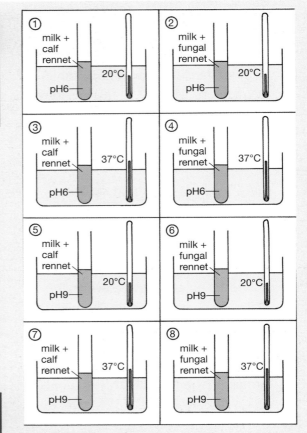

**Figure 5.32**

  (i) which pH makes calf rennet clot milk faster at 20°C;
  (ii) which pH makes fungal rennet clot milk faster at 37°C. (2)

**8** Table 5.13 shows the results from an experiment to investigate the effect of temperature on the action of calf rennet and fungal rennet.

a) Draw a line graph for each type of rennet on the same sheet of graph paper. Use two different colours and clearly label each line graph. (5)

b) At which temperature did calf rennet take the *least* time to make the milk clot? (1)

c) At which temperature did fungal rennet take the *least* time to make the milk clot? (1)

d) At which temperature does calf rennet work best? (1)

e) At which temperature does fungal rennet work best? (1)

| temperature (°C) | time for milk to clot (min) | |
|---|---|---|
| | calf rennet | fungal rennet |
| 10 | 50 | 52 |
| 15 | 46 | 46 |
| 20 | 40 | 38 |
| 25 | 32 | 30 |
| 30 | 20 | 20 |
| 35 | 10 | 5 |
| 40 | 5 | 10 |
| 45 | 20 | 30 |
| 50 | 56 | 60 |

**Table 5.13**

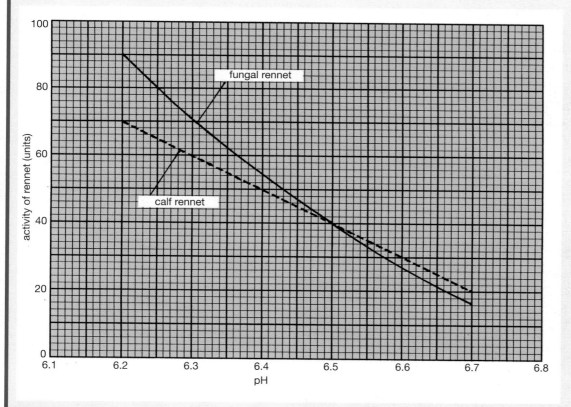

**Figure 5.33**

**9** Figure 5.33 shows a graph of the results from an experiment. It was set up to investigate the effect of pH on the activity of two types of rennet.

a) Which type of rennet was more active at pH 6.2? (1)

b) Which type of rennet was more active at pH 6.7? (1)

c) At which pH were the two types of rennet equally active? (1)

d) At which pH did calf rennet show an activity of 30 units? (1)

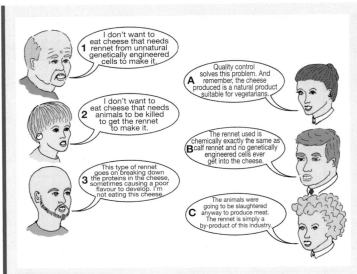

**Figure 5.34**

**10** Figure 5.34 shows three statements (1–3) made by members of the public concerned about types of rennet used in cheese-making. The diagram also shows the replies (A–C) given by three experts.

a) Match speech bubbles 1–3 with replies A–C and for each pair, state which type of rennet they are talking about. (3)

b) Which type of rennet are you most in favour of? Give ONE reason to support your choice. (1)

**11** When whey is released into a river, it affects the oxygen content of the river water. Figure 5.35 shows a graph of the results from tests carried out on a river at nine sample points.

a) What was the oxygen concentration of the river water before the whey was added? (1)

b) What effect did the addition of whey have on the river's oxygen concentration? (1)

c) At which sample site was the oxygen concentration of the water found to have returned to its starting level? (1)

e) How many units of activity were recorded for fungal rennet at pH 6.25? (1)

f) What was the difference in activity between the two types of rennet at pH 6.3? (1)

g) By how many times was the activity of calf rennet at pH 6.3 greater than that of calf rennet at pH 6.7? (1)

h) State the *range* of pH investigated in this experiment. (1)

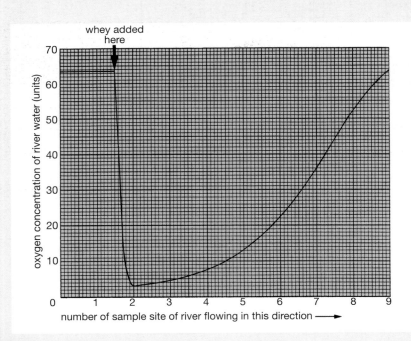

**Figure 5.35**

d) Which sample site's water would be found to contain the highest number of bacteria? (1)

e) By how many times is the oxygen concentration at sample site 7 greater than that at sample site 2? (1)

f) The sampling sites were 1 km apart. For how many km was the oxygen concentration below the starting level? (1)

g) Dumping whey in a river is now illegal in Britain. Suggest why. (1)

**12** A large dairy factory in Europe annually processes milk into 1250 tons of soft cheese, 850 tons of hard cheese, 700 tons of processed cheese and 300 tons of butter. For many years the whey left over from the process was poured into the local river. However a few years ago the management decided to try to sell the whey to farmers as part of feed suitable for sheep.

The farmers had been giving the sheep a sugar extract in their feed during the winter months. The dairy factory set up an experiment to persuade the sheep farmers that it would be to their advantage to add

whey instead of sugar to the animal feed. Table 5.14 shows the results of the experiment.

a) Construct a table to display the information about four milk products given in the first sentence of the passage. (2)

b) In the experiment, each group contained 80 animals. Why were so many used? (1)

c) Which group of animals was the *control* group? (1)

d) (i) On average, how much weight had an animal in group A gained after 64 days?

(ii) On average, how much weight had an animal in group B gained after 64 days?

(iii) On average, how much weight had an animal in group C gained after 64 days? (3)

e) From this experiment, which substance (whey or sugar extract) would you advise farmers to add to the sheep's winter food. Explain your choice. (2)

| group of sheep | substance added to animal feed | average weight of an animal (kg) | | average cost per animal for 64-day trial (€) |
| --- | --- | --- | --- | --- |
| | | at start | after 64 days | |
| A | water | 38 | 48 | 1 |
| B | whey | 38 | 52 | 10 |
| C | sugar extract | 38 | 52 | 30 |

**Table 5.14**

## What you should know

| | | |
|---|---|---|
| acid | flavour | rennet |
| animals | fungi | resazurin |
| bacteria | harmful | semi-skimmed |
| calves | organisms | skimmed |
| clot | oxygen | sugar |
| curds | pasteurised | UHT |
| engineered | preserving | upgraded |
| evaporated | proteins | whey |

**Table 5.15**  *Word bank for chapter 5*

1  Milk is a nourishing food that contains _____, fats, _____, vitamins and minerals.

2  Milk that has been treated to remove nearly all of its fat is described as _____. Milk that has had about half of its fat removed is described as _____. Milk that has been heated to remove much of its water, and preserve it, is described as _____.

3  Milk that has been heated to 72°C for 15 seconds to destroy disease-causing microbes is described as _____. Milk that has been heated to a very high temperature (e.g. 138°C) for a few seconds to destroy

microbes *and* bacteria which make milk go sour, is described as _____.

4  The presence of bacteria in milk can be demonstrated by the _____ dye test.

5  Yoghurt is produced from pasteurised milk by the action of useful _____. These convert sugar in milk to an _____ that thickens the milk. Making yoghurt is a way of _____ milk.

6  Cheese is produced from pasteurised milk by the action of _____ and useful bacteria.

7  Rennet makes the protein in milk clot as _____. The remaining liquid is called _____. Rennet can be obtained from _____, certain _____ and genetically _____ yeast cells.

8  Cheese-making bacteria _____ the milk protein and affect the _____ of the cheese.

9  Whey contains protein and sugar. If whey is poured into a river, the bacteria in the water feed on it and use up the water's _____ supply. This leads to a reduction in numbers and types of other _____.

10  Whey can be used to feed _____ or it can be _____ to some other useful product.

# 6 Yeast-based industries

## Yeast

Yeast is a simple fungus. Under the microscope it is seen to be made of single cells (see figure 6.1). When given a supply of food (e.g. sugar) and suitable growing conditions (e.g. 30°C and neutral pH), yeast cells multiply by budding (see figure 6.1).

### Growing yeast

Yeast can be grown in **liquid culture** in a fermenter (see page 138) or on jelly-like food called **nutrient agar**.

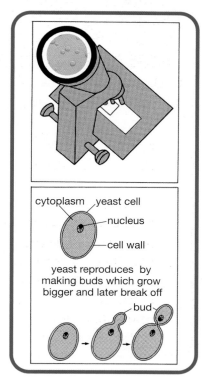

**Figure 6.1** *Yeast*

## More about aseptic conditions

During laboratory work with micro-organisms, certain precautions are taken to try to create **sterile (aseptic)** conditions in which to carry out the experiment. Some of these were described in chapter 5 (see page 108). Further precautions must be employed when preparing nutrient agar to grow microbes such as yeast.

### Preparation of sterile nutrient agar

**Agar powder** and **nutrients** such as sugar are added to pure water in a McCartney bottle and the cap replaced but not screwed on tightly. The bottle is placed in an **autoclave** (see figure 6.2). When this special container is sealed up and its heating element turned on, it can be

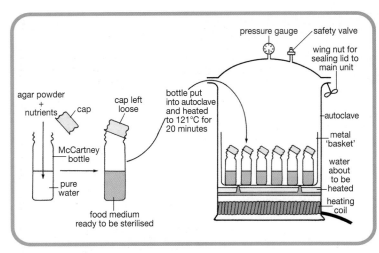

**Figure 6.2**   *Use of an autoclave*

heated up under pressure to temperatures well above the boiling point of water. Normally it is used to heat up equipment and agar to 121°C for 20 minutes since this treatment kills *all* microbes.

## Sterile Petri dish

Microbes are often cultured in containers called **Petri dishes** (see figure 6.3). These are also referred to as plates and may be made of plastic or glass. Glass Petri dishes can be sterilised in an autoclave. Plastic Petri dishes have already been sterilised during their manufacture. Their insides will stay sterile as long as they are not opened. A marker pen is normally used to write important information on the *base* of an unopened Petri dish.

## Preparing a plate of nutrient agar

Once the bottle of nutrient agar has been sterilised in the autoclave, it has to be poured into a sterile Petri dish to prepare a plate of nutrient agar. Further precautions must be taken at this stage to reduce the chance of a stray microbe from the air entering the nutrient agar.

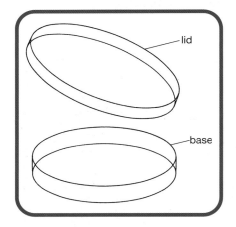

**Figure 6.3**   *A Petri dish*

- The work should be carried out close to a Bunsen flame so that the upward flow of warm air will carry any spores in the air up and away from the experiment.
- The cap of the McCartney bottle is removed and the bottle held in one hand while its neck is flamed for 2–3 seconds (see figure 6.4).
- The other hand is used to lift the lid of the Petri dish just enough to allow the nutrient agar to be quickly poured into the dish.
- The lid is immediately replaced and the dish is gently swirled once to spread out the molten agar before it sets.

Figure 6.5 shows a summary of the main precautions and sterile techniques carried out during laboratory work with micro-organisms.

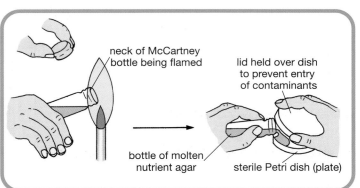

**Figure 6.4**   *Preparing a plate of nutrient agar*

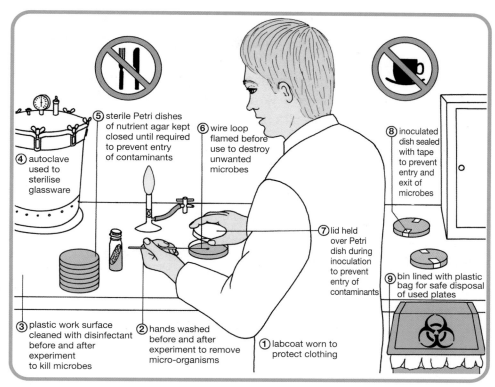

**Figure 6.5** *Summary of precautions and sterile techniques*

## Growing yeast on agar

You need
- Bunsen burner
- sterile plate of nutrient agar (suited to yeast's requirements)
- marker pen
- wire inoculating loop or sterile plastic inoculating loop
- sterilised McCartney bottle containing yeast suspension (dried yeast soaked in sterilised 1% glucose solution for one hour)
- access to discard jar of disinfectant if plastic inoculating loop is being used
- sellotape
- scissors
- access to incubator at 30°C

**Figure 6.6** *Spreading the yeast cells*

What to do
1 Wash your hands and work on a surface that has been cleaned with disinfectant.
2 Light the Bunsen burner.
3 Without opening your plate, turn it over base side up and write 'Yeast culture', the date and your initials near the edge.
4 Turn your plate back over ready for use.

5 If you are using a wire loop, sterilise it by flaming it in a Bunsen burner as shown in figure 5.11 on page 109. If you are using a plastic loop, carefully remove it from the cellophane wrapping by its handle end.

6 Remove the cap from the bottle of yeast suspension near the Bunsen and flame the neck of the bottle as shown in figure 5.12 on page 110.

7 Use your sterile loop to pick up a small sample of yeast cells.

8 Open the plate by lifting the lid just enough to allow the loop in, while at the same time keeping the lid over the base like a shield.

9 Using the loop, gently streak the plate in four lines as shown in figure 6.6.

10 Replace the lid as quickly as possible.

11 If you are using a wire loop, reflame it. If you are using a plastic loop, put it in the discard jar.

12 Use sellotape to seal your plate shut as shown in figure 6.7.

13 Incubate the plate at 30°C.

14 After 2–3 days, look at your plate but do not open it.

15 Make a large labelled diagram of your plate to show the results.

16 Hand in your plate for disposal by autoclaving.

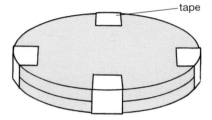

**Figure 6.7**   *Sealing a plate shut*

## Varieties of yeast

In large quantities, the yeast used in the above experiment appears light brown in colour; in tiny amounts it appears 'off-white'. Its colour may therefore be described as *brown* or *white*. A second type of yeast appears reddish pink in large quantities and pale pink in tiny amounts. Its colour may therefore be described as *pink* or *red*.

### Effect of temperature on the two types of yeast

Table 6.1 shows the results of growing brown and pink yeast at different temperatures. From these results it is concluded that pink

| colour of yeast | growth at 5°C | growth at 20°C | growth at 30°C | growth at 60°C |
|---|---|---|---|---|
| brown | ● | ●● | ●●● | ○ |
| pink | ●● | ●●● | ● | ○ |

key   ○    = no growth
      ●    = poor growth
      ●●   = good growth
      ●●●  = very good growth

**Table 6.1**   *Effect of temperature on growth of two types of yeast*

yeast grows best at around 20°C whereas brown yeast grows best at around 30°C. However the two types of yeast would need to be grown at many more temperatures (e.g.10°C, 15°C, 25°C and 35°C) to find out the temperature that produced the very best growth in each case.

# Use of yeast in bread dough

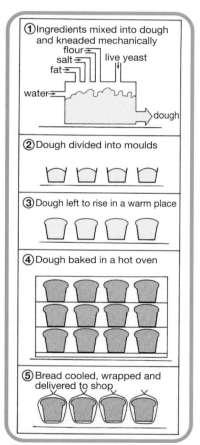

① Ingredients mixed into dough and kneaded mechanically
flour
salt
fat
live yeast
water
dough

② Dough divided into moulds

③ Dough left to rise in a warm place

④ Dough baked in a hot oven

⑤ Bread cooled, wrapped and delivered to shop

**Figure 6.8**  *Bread-making*

Yeast is a living organism. Like all other living things, it must **respire** 24 hours a day to release the energy needed to stay alive. When yeast is in its dry (dormant) state, its rate of respiration is very low. When yeast is given suitable growing conditions it becomes active. Its respiration rate increases and it releases much **carbon dioxide** ($CO_2$) gas.

If actively respiring yeast cells are used as one of the ingredients in bread dough, the bubbles of carbon dioxide gas released become caught in the sticky dough and make it **rise**. The dough is then baked in an oven at a high temperature. This turns the dough into bread and kills the yeast cells. Figure 6.8 shows the sequence of events involved in making bread on a commercial scale.

## Activity

### Investigating how successful different types of yeast are at raising dough

You need
- six labels
- three beakers (250 cm³ plastic)
- three measuring cylinders (100 cm³ plastic)
- teaspoon (5 cm³)
- flour
- glucose powder

# Biotechnological Industries

- three weigh boats
- fresh yeast
- dried yeast
- dead yeast
- access to electronic balance
- measuring cylinder (10 cm³ plastic)
- three stirring rods
- access to incubator at 30°C

## What to do (also see figure 6.9)

1. Make a copy of table 6.2.
2. Label the beakers A, B and C and the 100cm³ measuring cylinders A, B and C.
3. Add your initials to the labels on the measuring cylinders.
4. Add 10 heaped teaspoonfuls of flour and one heaped teaspoonful of glucose to each beaker.
5. Weigh out 1 g of fresh yeast in a weigh boat and add it to beaker A.
6. Weight out 1 g of dried yeast in a weigh boat and add it to beaker B.
7. Weigh out 1 g of dead yeast in a weigh boat and add it to beaker C.
8. Add 40 cm³ of water to each beaker using the small plastic measuring cylinder.
9. Stir each beaker equally using a separate stirring rod.
10. Carefully transfer the contents of each beaker into the appropriate measuring cylinder.
11. Record the volume of 'dough' in each measuring cylinder at the start in your table.
12. Leave the three measuring cylinders in the incubator at 30°C.
13. After 24 hours record the volume of dough in each cylinder in your table.
14. Complete the table.
15. Draw a conclusion about how successful different types of yeast are at making dough rise.

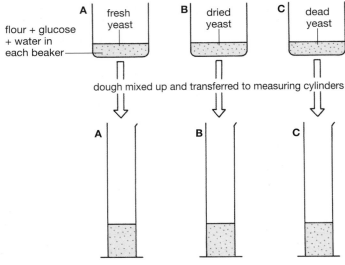

**Figure 6.9** *Using yeast to raise dough*

| | | type of yeast | |
| --- | --- | --- | --- |
| | fresh (A) | dried (B) | dead (C) |
| volume of 'dough' at start (cm³) | | | |
| volume of dough after 24 hours (cm³) | | | |
| change in volume of dough (cm³) | | | |

**Table 6.2** *Results for different yeasts in dough*

## Use of yeast in beer-making

### Activity

## Investigating the action of yeast on sugar in a liquid culture

You need
- two labels
- two conical flasks (100 cm³)
- measuring cylinder (100cm³ plastic)
- 200 cm³ of boiled and cooled glucose solution (1%)
- weigh boat containing 1 g dried yeast
- weigh boat containing 1 g dead yeast powder
- plastic dropper
- lime water
- two fermentation locks

What to do
1 Label the conical flasks A and B and add your initials.
2 Using the measuring cylinder, add 100 cm³ of boiled and cooled glucose solution to each conical flask.
3 Add 1 g of dried yeast to flask A and 1 g of dead yeast to flask B.
4 Using the dropper, add lime water to the U-bend in each fermentation lock.
5 Fit the fermentation locks to the conical flasks.
6 See figure 6.10 which shows flask A set up at the start of the experiment. (Note: Flask B is the **control**. It is exactly the same as flask A except that it contains dead yeast instead of live yeast. This allows a fair comparison of the results to be made and a conclusion to be drawn about the effect of the one variable factor.)
7 After 30 minutes, check flasks A and B for bubbles and the lime water for any changes.
8 Repeat step 7 after 1–2 days.
9 Unplug the rubber stopper in each flask and smell the contents for fumes that indicate that alcohol has been formed.
10 Draw a table of your results for both the lime water test and the smell test.
11 Draw two conclusions about the action of yeast on sugar.

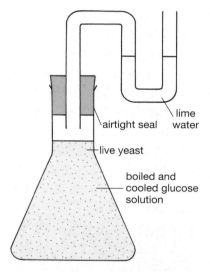

airtight seal
lime water
live yeast
boiled and cooled glucose solution

**Figure 6.10** *Investigating the action of yeast*

## Beer

**Figure 6.11** *Fermenter*

The process by which yeast converts sugar into alcohol and carbon dioxide is called **fermentation**. It can be represented by the following equation:

$$\text{sugar} \xrightarrow{\text{yeast}} \text{alcohol} + \text{carbon dioxide}$$

A **fermenter** is a container in which a fermentation process is brought about by a microbe such as yeast. The flask in figure 6.10 is a simple fermenter.

**Beer** is an alcoholic drink. It is brewed in enormous fermenters (see figure 6.11) from malt, water, sugar, hops and yeast. Figure 6.12 shows the main processes involved in the commercial brewing of beer.

## Alcohol content of beers

Many different types of beer exist. A few examples are given in table 6.3. On average the **alcohol content** of beer is about 3–4% (see figure 6.13). However it varies depending on:

- the type of yeast used during brewing;
- the temperature at which the fermenter is kept;
- the length of the fermentation time.

**Figure 6.13** *Alcohol content of beers*

| type of beer | alcohol (%) |
|---|---|
| alcohol-free beer | 0.05 |
| low alcohol beer | 1.1 |
| bitter | 3.0 |
| heavy | 3.7 |
| lager | 4.0 |
| stout | 4.2 |
| export ale | 4.5 |
| super lager | 8.5 |
| barley ale | 11.0 |

**Table 6.3** *Alcohol content of beers*

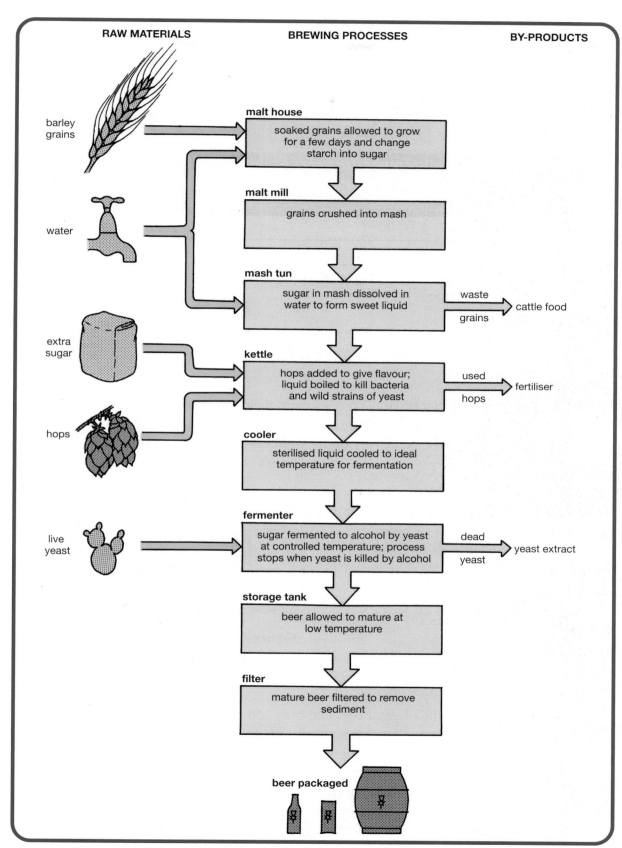

**Figure 6.12** *Commercial brewing of beer*

## Top- and bottom-fermenting yeasts

**Top-fermenting** yeasts rise to the top of the fermentation vessel at the end of the fermentation time, whereas **bottom-fermenting** yeasts sink to the bottom. These two types of yeast also differ in temperature requirements and length of fermentation time as summarised in table 6.4.

| type of yeast | temperature range required (°C) | length of fermentation time (days) | type of beer produced |
|---|---|---|---|
| top-fermenting | 14–20 | 6 | ale |
| bottom-fermenting | 8–14 | 20 | lager |

**Table 6.4**  *Comparison of two beer-making yeasts*

## Activity

### Investigating the effect of temperature on the activity of yeast

#### Information
When live yeast cells are placed in glucose solution, their rate of respiration increases and they release bubbles of carbon dioxide. This gas gathers as a **froth of bubbles** above the yeast culture. The larger the volume of froth, the greater the activity of the yeast cells.

#### You need
- three labels
- three measuring cylinders (100 cm³ glass)
- dropping bottle of yeast suspension (e.g. fast action Hovis or Allinson)
- dropping bottle of glucose solution
- measuring cylinder (10 cm³ plastic)
- access to water baths at 25°C, 35°C and 45°C

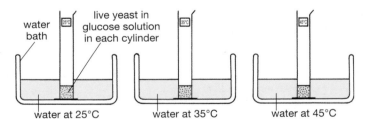

**Figure 6.14**  *Investigating the effect of temperature on yeast activity*

#### What to do (also see figure 6.14)
1 Label the three measuring cylinders 25°C, 35°C and 45°C and add your initials.
2 Shake the dropping bottle of yeast cells and then add 10 cm³ to each measuring cylinder.
3 Using the small measuring cylinder, add 10 cm³ of glucose solution to each measuring cylinder.
4 Place the three glass measuring cylinders in the appropriate water baths.
5 Prepare your table of results as shown in table 6.5.
6 After 20 minutes, collect the three measuring cylinders and record the results in your table.
7 If other groups have done the same experiment, pool your results.
8 Draw a line graph of the results with temperature on the horizontal (x) axis and average volume of froth on the vertical (y) axis.
9 Draw a conclusion about the effect of temperature on the activity of yeast.

| | temperature (°C) | | |
|---|---|---|---|
| | 25 | 35 | 45 |
| volume of yeast culture at start (cm³) | | | |
| volume of yeast culture and froth after 20 min (cm³) | | | |
| my group's result for volume of froth (cm³) | | | |
| another group's result for volume of froth (cm³) | | | |
| average volume of froth (cm³) | | | |

**Table 6.5** *Results for yeast investigation 1*

## Activity

## Investigating the activity of different types of yeast

You need
- two labels
- two measuring cylinders (100 cm³ glass)
- dropping bottle of brewer's yeast suspension
- measuring cylinder (10 cm³ plastic)
- dropping bottle of baker's yeast suspension (e.g. fast action Hovis or Allinson)
- dropping bottle of glucose solution (1%)
- access to water bath at 35°C

What to do (also see figure 6.15)
1 Label the glass measuring cylinders A and B and add your initials.
2 Shake the dropping bottle of brewer's yeast suspension and use the small measuring cylinder to add 10 cm³ to measuring cylinder A.

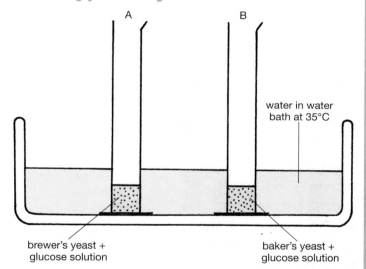

water in water bath at 35°C

brewer's yeast + glucose solution

baker's yeast + glucose solution

**Figure 6.15** *Investigating the activity of different types of yeast*

3 Rinse out the small measuring cylinder and then repeat step 2 for baker's yeast in measuring cylinder B.
4 Rinse out the small measuring cylinder and use it to add 10 cm³ of glucose solution to measuring cylinders A and B.
5 Place cylinders A and B in the water bath at 35°C.
6 Prepare your results table as shown in table 6.6.
7 After 20 minutes, collect the two cylinders and record the results in your table.
8 If other groups have done the same experiment, pool your results.
9 Draw a bar graph of the results with average volume of froth on the vertical (y) axis.
10 Draw a conclusion about the activity of the two types of yeast.

| | type of yeast | |
| --- | --- | --- |
| | brewer's in cylinder A | baker's in cylinder B |
| volume of yeast culture at start (cm³) | | |
| volume of yeast culture and froth after 20 min (cm³) | | |
| my group's result for volume of froth (cm³) | | |
| another group's result for volume of froth (cm³) | | |
| average volume of froth (cm³) | | |

**Table 6.6**   *Results for yeast investigation 2*

# Maturing the beer

**Figure 6.16**   *Real ale*

Following its time in the fermenter (see figure 6.12) the beer is allowed to **mature** in a storage container to improve its flavour. There are two ways of storing the beer as follows.

## 1 Brewery-conditioned beer

This is the type of beer brewed in figure 6.12. It is stored in large tanks. The yeast cells are removed from it and additional carbon dioxide is added before it is packaged in bottles, cans and kegs ready for sale.

## 2 Cask-conditioned beer

This type of beer is stored in huge wooden (or sometimes steel) containers called **casks**. The yeast cells are not removed. Instead they are given more sugar to encourage them to keep growing and producing carbon dioxide. The beer formed is normally dark in colour, highly flavoured and known as **real ale** (see figure 6.16).

## Testing your knowledge

**1** Rewrite the following two sentences to include only the correct answer from each choice given in italics.

Yeast is a simple *fungus/bacterium*. It is made up of many *cells joined together in chains/single cells*. (2)

**2** a) What is meant by the term *aseptic* conditions? (1)
   b) (i)   What is the name of the jelly-like food material upon which micro-organisms are often grown?
      (ii)  By what means can a bottle of this food material be sterilised before use?
      (iii) Describe the correct procedure for pouring this hot jelly-like food into a Petri dish. (5)
   c) (i)   State how you would sterilise a metal inoculating loop.
      (ii)  Describe how you would use the loop to inoculate a Petri dish with yeast cells from a culture bottle. (5)

**3** a) (i)   What gas is released by live yeast cells?
      (ii)  Why is this useful in bread- making? (1)
   b) Which part of the bread-making process ensures that the yeast cells in bread are dead? (1)

**4** a) Give a simple word equation that represents the process of fermentation. (2)
   b) State TWO factors that can affect the alcohol content of beer. (2)

**5** Decide whether each of the following statements is true or false and then use T or F to indicate your choice. Where a statement is false, give the word that should have been used in place of the word(s) in **bold** print. (9)
   a) Alcohol-free beer contains a **tiny quantity** of alcohol.
   b) Super lager contains **less** alcohol than export ale.
   c) Top-fermenting yeast works best at **14–20°C**.
   d) Top-fermenting yeast produces ale after **20** days.
   e) Bottom-fermenting yeast works best at **14–20°C**.
   f) Bottom-fermenting yeast produces **lager** after 20 days.
   g) Extra carbon dioxide is added to **brewery**-conditioned beer before packaging.
   h) Extra sugar is added to **cask**-conditioned beer to keep the yeast growing.
   i) **Brewery**-conditioned beer is also known as real ale.

## Enzymes

**Enzymes** are substances produced by all living cells. Each enzyme controls a particular **biochemical reaction** (see figure 7.1 on page 162). An enzyme can be extracted from cells and used to control the same biochemical reaction on a large scale in a fermenter as part of an industrial process.

### Fermented milk drink

**Kefir** is an alcoholic fermented milk drink. It has been produced for many generations by local people in Eastern European countries. It is

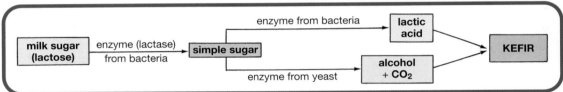

**Figure 6.17** *Production of kefir*

made by adding a colony (called a 'grain') of useful **bacteria** and **yeast** cells to milk. Enzymes made by these living cells promote the series of chemical reactions shown in figure 6.17.

The *lactic acid* gives the kefir a mildly sour taste and the *carbon dioxide* makes it slightly fizzy. The *alcohol* content of kefir is usually about 2%.

### Immobilisation

It is possible to make a type of fermented milk drink by adding **lactase** enzyme, live **yeast** cells and some live **yoghurt** to whole milk. However it is difficult to extract the lactase enzyme needed for this process from bacteria. In addition it is wasteful to throw away the enzyme and the yeast after use. Fortunately scientists have found a way of trapping valuable resources like these so that they can be reused. The trapping process is called **immobilisation**.

In the examples that follow, an **immobilised** enzyme or cell is one that cannot move freely because it has been deliberately trapped in **pellets of gel** (see figure 6.18). The enzyme or cell (immobilised on the surface of the gel pellet) is then put to work. After it has brought about the required chemical reaction, it can be easily separated from the end product and *used again*.

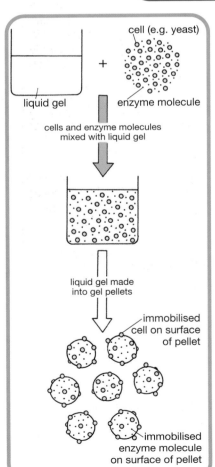

**Figure 6.18** *Immobilisation*

## Activity

# Making gel pellets containing immobilised enzyme and yeast cells

### You need

- 2 cm³ lactase enzyme in a 10 cm³ measuring cylinder
- 2 cm³ water containing wine-making yeast in a 100 cm³ plastic beaker
- 8 ml sodium alginate solution and one drop of food colouring in a 10 cm³ syringe
- stirring rod
- 50 cm³ calcium chloride solution in a 100 cm³ conical flask
- label
- tea strainer
- metal foil

### What to do (also see figure 6.19)

1. Pour the lactase enzyme into the beaker containing the yeast suspension.
2. Add the coloured sodium alginate from the syringe to the lactase and yeast.
3. Stir the mixture thoroughly using the stirring rod.
4. Draw up the mixture in the syringe.
5. Carefully add the contents of the syringe, one drop at a time, to the calcium chloride solution in the conical flask.
6. Leave the gel pellets in the calcium chloride solution for 10 minutes to let them set.
7. During this time, wash out the plastic beaker and add a label to it with your initials.
8. Empty the pellets into the tea strainer at the sink and rinse them with cold water.
9. Transfer the pellets to the labelled plastic beaker and add enough cold water to cover them.
10. Add a lid of metal foil to the beaker and store it in the fridge until the next lesson.

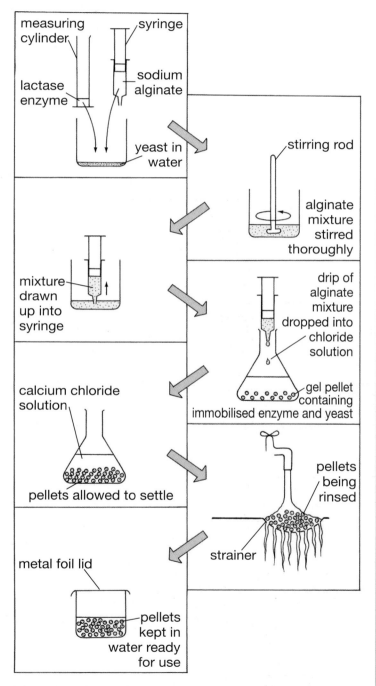

**Figure 6.19** *Making gel pellets containing immobilised enzyme and yeast*

## Activity

### Making a fermented drink using immobilisation techniques

You need
- spatula
- natural 'live' yoghurt
- $100\,cm^3$ whole milk in a $250\,cm^3$ glass beaker
- two plastic droppers
- two test tubes
- test tube stand
- two Clinistix strips
- dropping bottle of pH indicator solution
- pH reference strip
- supply of gel pellets containing immobilised lactase enzyme and yeast cells
- tea strainer
- metal foil
- empty $250\,cm^3$ glass beaker
- access to water bath at 38°C

What to do (also see figure 6.20)
1 Prepare a copy of results table 6.7.
2 Using a spatula, add a scoop of natural 'live' yoghurt to the milk and mix thoroughly.
3 Using a plastic dropper, transfer a small sample of the milk–yoghurt mixture to each test tube.
4 Test the first one for simple sugar (glucose) using a Clinistix strip. (Positive result = pink to purple in less than 10 seconds.)
5 Add three drops of universal indicator to the second sample and find out the starting pH of the milk–yoghurt mixture.
6 Complete the 'at start' column in your results table.
7 Pour the gel pellets into the tea strainer at the sink and then add them to the milk–yoghurt mixture.
8 Cover the beaker with a lid of metal foil and place it in the water bath at 38°C.
9 After 4–6 hours collect your beaker, lift the lid and smell the drink for signs of alcohol fermentation.
10 Separate the gel pellets from the drink using the tea strainer held over the second glass beaker. Look at the drink's texture and watch for signs of bubbles indicating release of carbon dioxide.

| test or observation | state of fermented milk drink | |
| --- | --- | --- |
| | at start | after 4–6 hours at 38°C |
| 1 Does Clinistix test show presence of simple sugar (glucose)? | | |
| 2 What is the pH? | | |
| 3 Are there signs of $CO_2$ bubbles? | | |
| 4 Is there a smell indicating alcohol fermentation? | | |
| 5 Is the texture runny or creamy? | | |

**Table 6.7** *Fermented milk drink results*

11 Using a fresh dropper, take two samples of the fermented drink and repeat steps 4 and 5.
12 a) Complete your table to show five ways in which the fermented milk drink differs from the original milk–yoghurt mixture.
   b) With reference to the chemical reactions given in figure 6.17, try to give a reason for each of these differences.
13 Rinse the pellets in the small plastic beaker and reuse them by repeating the experiment with a fresh supply of the milk–yoghurt mixture.

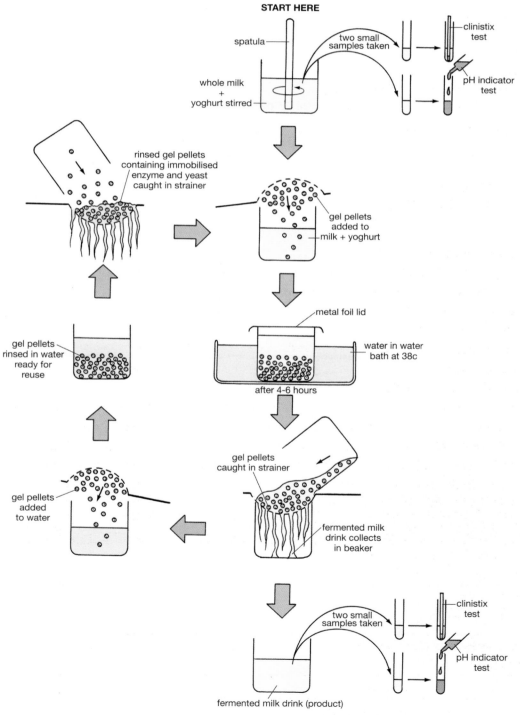

**Figure 6.20**   *Making a fermented milk drink*

## Immobilising yeast cells (assessment)

You need

- 8 cm³ sodium alginate solution containing one drop of food colouring in a 10 cm³ syringe
- 2 cm³ of live yeast suspension in a 100 cm³ plastic beaker
- stirring rod
- 50 cm³ of calcium chloride solution in a 100 cm³ conical flask
- tea strainer

What to do (also see figure 6.21)

1 Add the coloured sodium alginate from the syringe to the yeast suspension.
2 Stir the mixture thoroughly using the stirring rod.
3 Draw up the mixture in the syringe.
4 Carefully add the contents of the syringe, one drop at a time, to the calcium chloride solution.
5 Leave the gel pellets in the calcium chloride solution for 10 minutes to set.
6 During this time, wash out the plastic beaker.
7 Empty the pellets into the tea strainer at the sink and rinse them with cold water.
8 Transfer the pellets to the plastic beaker and show them to the teacher to have your work assessed.

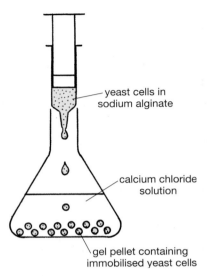

yeast cells in sodium alginate

calcium chloride solution

gel pellet containing immobilised yeast cells

**Figure 6.21** *Immobilising yeast cells*

## Food flavouring

**Figure 6.22** *Foods flavoured with yeast*

A **flavouring** is a substance used to add to the taste and/or smell of a food. Yeast is used to flavour foods such as soups and snacks. Stock cubes and powder mixes used to make soups, casseroles, sauces and gravies often contain yeast extract (see figure 6.22).

## Activity

# Demonstrating flavour development in yeast

### You need
- teaspoon
- salt
- beaker (250 cm³ plastic)
- sterile glass rod
- block of fresh yeast (see figure 6.23)
- sterile Petri dish (glass)
- access to incubator at 45°C

### What to do (also see figure 6.24)
1 Prepare a copy of results table 6.8.
2 Add three teaspoonfuls of salt to the block of fresh yeast in the beaker.
3 Using the glass rod, mix the yeast and salt thoroughly until a 'wet' mixture is formed.
4 Pour the 'wet' mixture into the base of the Petri dish, smell the mixture and note the result in your table.
5 Add the lid and place the dish in the incubator at 45°C.
6 After 12–24 hours, gently lift the lid and immediately smell the contents of the dish. (Do *not* taste them.)
7 Complete your table to show the results.
8 Draw a conclusion from the experiment.

**Figure 6.23** *Blocks of fresh yeast*

| time | appearance of yeast–salt mixture | smell of yeast–salt mixture |
|---|---|---|
| after stirring but before heat treatment | | |
| after 12–24 hours at 45°C | | |

**Table 6.8** *Flavour development results*

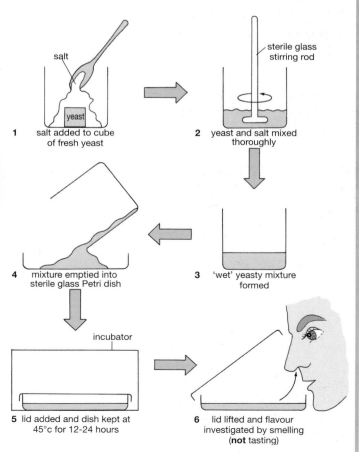

1 salt added to cube of fresh yeast

2 yeast and salt mixed thoroughly

3 'wet' yeasty mixture formed

4 mixture emptied into sterile glass Petri dish

5 lid added and dish kept at 45°c for 12-24 hours

6 lid lifted and flavour investigated by smelling (**not** tasting)

salt

yeast

sterile glass stirring rod

incubator

**Figure 6.24** *Flavour development in yeast*

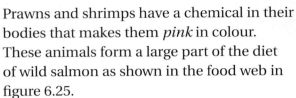
## Food colouring

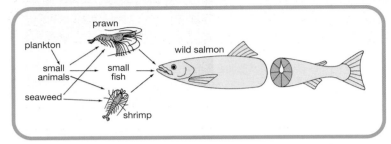

**Figure 6.25**  *Food web*

Prawns and shrimps have a chemical in their bodies that makes them *pink* in colour. These animals form a large part of the diet of wild salmon as shown in the food web in figure 6.25.

Once inside the salmon's body, the pink substance makes its flesh pink. However salmon reared in fish farms are not fed pink prawns and shrimps and their flesh is found to be a *dull grey* colour. This makes the fish 'steaks' unattractive and difficult to sell. Fish farmers have overcome this problem by adding red dye to the salmon's diet just before they are killed. The dye makes the salmon's flesh turn pink but does not affect the taste of the fish. Some people have criticised this practice. They claim that it makes people consume an unnatural chemical.

### Pink yeast

It is now known that a pink variety of yeast contains the same pink substance as that found in shrimps and prawns. It is possible to make the farmed salmon's flesh turn pink by mixing this pink yeast with the normal fish food that they are given. Many people regard this as a preferable method of colouring the salmon's flesh since the pink colour has come from a *natural* source.

### Fish foods

A wide variety of fish foods are produced for use in fish farms and aquaria. Many of these, such as tropical flakes, *Spirulina* sticks, grow crumbles and goldfish flakes (see figure 6.26), contain yeast as one of their main ingredients (see table 6.9).

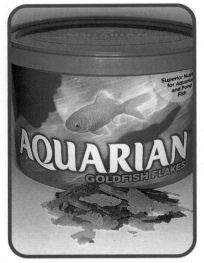

**Figure 6.26**  *Fish food containing yeast*

| ingredient | fish food | | | | | | |
|---|---|---|---|---|---|---|---|
| | brine-shrimp flakes | egg flakes | tropical flakes | Spirulina sticks | growth crumbles | earth-worm sticks | goldfish flakes |
| white fish meal | ✓ | ✓ | ✓ | ✓ | ✓ | ✓ | ✓ |
| soy flour | ✓ | ✓ | ✓ | ✓ | ✓ | ✓ | ✓ |
| yeast | ✓ | ✓ | ✓ | ✓ | ✓ | ✓ | ✓ |
| plankton | ✓ | ✗ | ✓ | ✓ | ✓ | ✓ | ✗ |
| brine-shrimps | ✓ | ✓ | ✓ | ✓ | ✓ | ✓ | ✗ |
| earthworms | ✗ | ✗ | ✗ | ✗ | ✓ | ✓ | ✗ |
| *Spirulina* | ✓ | ✗ | ✓ | ✓ | ✓ | ✓ | ✓ |
| vitamins | ✓ | ✓ | ✓ | ✓ | ✓ | ✓ | ✓ |
| dried egg | ✗ | ✓ | ✗ | ✗ | ✗ | ✗ | ✓ |
| krill | ✗ | ✗ | ✓ | ✓ | ✓ | ✓ | ✗ |
| kelp | ✗ | ✗ | ✓ | ✓ | ✓ | ✓ | ✗ |

**key** ✓ = present ✗ = absent

**Table 6.9**  *Typical ingredients of fish foods*

# Environmental impact

## Potential effect of waste from yeast-based industry

If waste from yeast-based industries (such as brewing) is released into a river, it can have a similar effect to that of whey (see page 121). The waste acts as a source of food for bacteria already present in low numbers in the river water. These bacteria increase rapidly in number and use up most of the river water's dissolved **oxygen** supply. River animals such as fish suffer and may even die from lack of oxygen.

### Oxygen demand
Polluted river water that has had most of its oxygen used up by microbes is said to have an **oxygen demand**. The more oxygen-consuming microbes that it contains, the higher its oxygen demand.

## Methylene blue dye

**Methylene blue dye** is a chemical that stays blue when added to a sample of water that contains plenty of dissolved oxygen (e.g. cold tap water). However its blue colour turns to colourless in water that contains very little oxygen. When this happens it shows that the water has a high oxygen demand.

### Activity

# Investigating the oxygen demand of water contaminated with yeast

You need

- two labels
- two test tubes
- test tube stand
- measuring cylinder (10 cm³ plastic)
- suspension of live yeast cells that have been aerated in 1% glucose solution for an hour
- suspension of dead yeast cells in 1% glucose solution
- dropping bottle of methylene blue dye (1%)
- two rubber stoppers
- access to water bath or incubator at 40°C

What to do (also see figure 6.27)

1 Prepare a copy of results table 6.10.
2 Label the test tubes A and B and add your initials.
3 Using the measuring cylinder, add 10 cm³ of live yeast suspension to tube A.
4 Rinse out the measuring cylinder and use it to add 10 cm³ of dead yeast suspension to tube B.
5 Add two drops of methylene blue dye to each tube.
6 Fit the stoppers and invert each tube once to mix the contents.
7 Place the tubes in the water bath or incubator at 40°C.
8 After 30 minutes, inspect your tubes and enter the results in the table.
9 Draw a conclusion about the oxygen demand of water contaminated with live yeast compared with that of water containing dead yeast cells.
10 Shake up tube A vigorously. Suggest why the blue colour reappears.
11 Which tube was the control? Why was a control set up in this experiment?

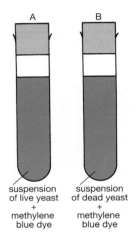

suspension of live yeast + methylene blue dye

suspension of dead yeast + methylene blue dye

**Figure 6.27**  *Methylene blue dye experiment*

| time | colour of contents of tube | |
| --- | --- | --- |
| | **A** | **B** |
| at start | blue | blue |
| after 30 min | | |

**Table 6.10**  *Results of methylene blue dye experiment*

## Upgrading and use of waste

Yeast-based industries no longer release their wastes into the local river. Instead of regarding this material as useless, they upgrade it. **Upgrading** means changing a waste material into a useful product. For example, spent barley grains and yeast cells from the brewing industry are used to produce animal feed such as **cattle cake**. Yeast is also involved in the upgrading of 'waste' **whey** (see figure 5.25 on page 122).

## Testing your knowledge

**1** a) Where are enzymes produced naturally? (1)
b) What does an enzyme do? (1)

**2** a) (i) Name TWO types of living organism needed to make a fermented milk drink.
(ii) Which of these makes an enzyme that breaks down the sugar in milk to a simpler form?
(iii) Which of these makes an enzyme that changes this simple sugar into alcohol? (4)
b) Why does a fermented milk drink taste slightly sour? (1)
c) What gas is released during the fermentation process? (1)

**3** a) When an enzyme or cell is described as being *immobilised*, what does this mean? (2)
b) Describe how live yeast cells could be immobilised in the science lab. (4)
c) What is the advantage of immobilising a type of cell or enzyme? (1)

**4** Decide whether each of the following is true or false and then use T or F to indicate your choice. Where a statement is false, give the word that should have been used in place of the word(s) in **bold** print. (9)
a) Yeast is used in foods such as soups and sauce mixes to add **colouring**.
b) The flesh of wild salmon is pink because they feed on pink **shrimps**.
c) If left untreated, the flesh of salmon in fish farms is **dull grey**.
d) The flesh of salmon in fish farms can be made to go pink by adding cells of red **bacteria** to their food.
e) Waste from yeast-based industry can have a **helpful** effect on a river.
f) The yeast wastes are used as food by river **bacteria**.
g) The bacteria use up the river's **carbon dioxide** supply.
h) Upgrading means changing a waste material into a **useless** product.
i) Wastes from yeast-based industries can be used to make **cattle cake**.

## Applying your knowledge

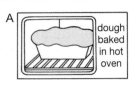

A
dough baked in hot oven

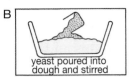

B
yeast poured into dough and stirred

C
freshly baked bread

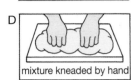

D
mixture kneaded by hand

E
yeast releases $CO_2$ bubbles making dough rise

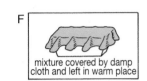

F
mixture covered by damp cloth and left in warm place

**Figure 6.28**

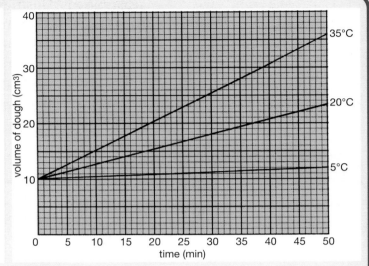

**Figure 6.29**

1 Figure 6.28 shows six stages (A–F) involved in bread-making. Arrange them into the correct sequence. (1)

2 A student set up an experiment to investigate the effect of temperature on the action of baker's yeast. He made the dough with yeast and used it to fill three measuring cylinders up to the 10 cm³ mark. He then left the cylinders at 5°C, 20°C and 35°C for 50 minutes. The graph in figure 6.29 shows his results.

a) (i) What was the one variable factor studied in this experiment?
 (ii) How many different values of this variable factor were used? (2)
b) What effect did an increase in temperature have on the volume of the dough produced? (1)
c) What was the volume of dough at 35°C after 35 minutes? (1)
d) At which temperature was the volume of dough found to be 11 cm³ after 25 minutes? (1)
e) How long did it take dough at 20°C to reach a volume of 22 cm³? (1)

f) By how many times was the volume of dough at 35°C greater than at 5°C after 50 minutes? (1)

3 Figure 6.30 shows an experiment set up to investigate respiration in yeast.
a) (i) Name the gas that has gathered in the test tube making it float.
 (ii) What test is used to identify this gas? (2)
b) Draw a labelled diagram of the control experiment that should have been set up to show that the gas was released by the yeast cells. (2)

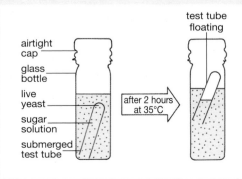

airtight cap
glass bottle
live yeast
sugar solution
submerged test tube

after 2 hours at 35°C

test tube floating

**Figure 6.30**

c) What apparatus could be used to keep the bottle at 35°C during the experiment? (1)

d) Predict the result of the experiment if it had been carried out at
   (i) 10°C,
   (ii) 70°C.
   (iii) Explain your answers to (i) and (ii). (4)

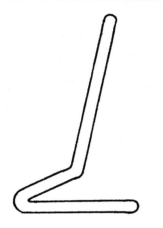

**Figure 6.31**

alcohol allowed to burn off

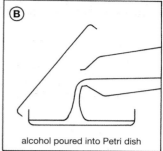
alcohol poured into Petri dish

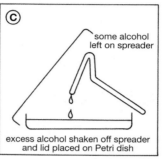
some alcohol left on spreader

excess alcohol shaken off spreader and lid placed on Petri dish

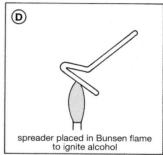
spreader placed in Bunsen flame to ignite alcohol

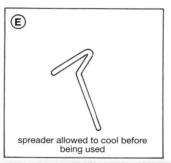
spreader allowed to cool before being used

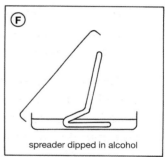
spreader dipped in alcohol

**Figure 6.32**

4 Figure 6.31 shows a glass spreader. This tool is often used to spread micro-organisms over the surface of a dish of nutrient agar. But before it can be used, it must first be sterilised. Figure 6.32 shows six steps (A–F) carried out to sterilise a spreader using burning alcohol. Arrange them into the correct order starting with step B. (1)

5 Figure 6.33 shows apparatus A set up to find out if live yeast cells can ferment sucrose sugar to alcohol. Apparatus B was set up as the control. State FIVE ways in which B needs to be changed to make it a valid control. (5)

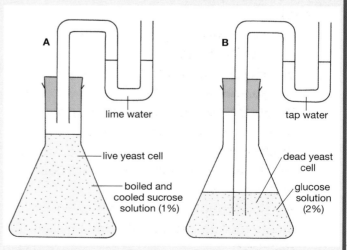

A  lime water — live yeast cell — boiled and cooled sucrose solution (1%)

B  tap water — dead yeast cell — glucose solution (2%)

**Figure 6.33**

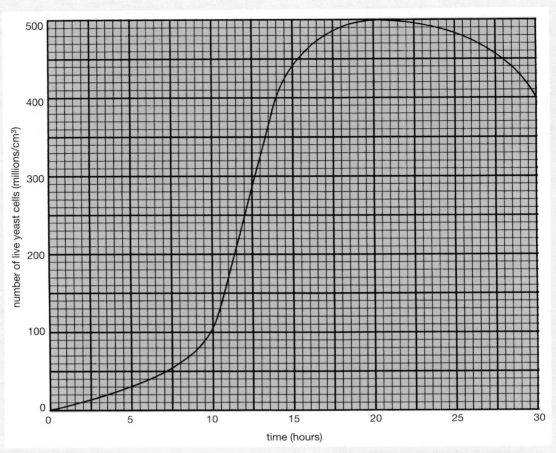

**Figure 6.34**

6 The graph in figure 6.34 shows changes in the number of live yeast cells growing in liquid culture at 32°C for 30 hours.
   a) (i)   How many yeast cells were present per cm³ at 5 hours?
      (ii)  From this point onwards, how many more hours passed before the yeast had doubled in number? (2)
   b) During which of the following periods of time did the yeast cells show the greatest increase in number?
      **A** 0–5 hours **B** 5–10 hours **C** 10–15 hours **D** 15–20 hours (1)
   c) For how many hours was the number of yeast cells found to be above 400 million per cm³? (1)
   d) Suggest why the number of yeast cells began to decrease after 23 hours. (1)

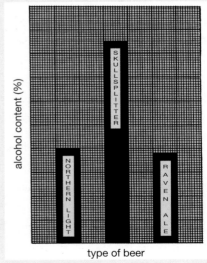

type of beer

**Figure 6.35**

7 The bar graph in figure 6.35 represents the alcohol content of three types of cask-conditioned beer made by Orkney Brewery.
   a) Make a copy of the bar graph on similar graph paper. (1)

b) *Northern Light* beer contains 4% alcohol. Use this information to add a scale to the vertical (y) axis. (1)

c) State the alcohol content of
   (i) *Skullsplitter* beer
   (ii) *Raven Ale*. (2)

d) A fourth beer called *Dark Island* contains 4.6% alcohol. Add a bar to your graph to include it. (1)

e) By how many per cent is the alcohol content of *Skullsplitter* greater than that of *Raven Ale*? (1)

8 Read the passage and answer the questions.
   a) (i) Which fossil fuel is most commonly used to run cars in Britain?
   (ii) Why is this fuel not used by all motorists in Brazil? (2)

b) Approximately how many tons of Brazil's annual sugar cane crop are used to make gasohol? (1)

c) Give the word equation of fermentation. (2)

d) Explain why yeast is not able to make sugar change into a liquid that is 100% alcohol. (2)

e) Why is alcohol described as a renewable resource? (2)

9 Different types of yeast are used to make different types of beer. Table 6.11 shows the results from an experiment set up to investigate the effect of temperature on two types of yeast. The volume of carbon dioxide released was used as a measure of yeast activity.

## GASOHOL

Brazil does not have rich supplies of fossil fuels such as coal, oil and gas. However it does have a warm, sunny climate and plenty of land. It is therefore able to grow sugar cane on a large scale. Sugar cane plants make a lot of sugar. If yeast is added to this sugar, the sugar is changed into alcohol by fermentation. Carbon dioxide is released.

Alcohol is poison. Most of the yeast cells die when the concentration of the culture liquid reaches 12–15% alcohol. Since alcohol boils at 80°C, it can be separated from the watery fermentation mixture by distillation (see figure 6.36).

About half of Brazil's 200 million ton sugar cane crop is converted into alcohol every year. The alcohol is then used as fuel for cars. The product on sale at the pumps is called **gasohol** (see figure 6.37). It is mostly alcohol but also contains a little petrol to stop people drinking it.

Fossil fuels like oil are non-renewable; once they run out, they have gone forever. Alcohol, on the other hand, is a renewable resource. As long as the Sun shines and plants make sugar, the sugar can be made into alcohol.

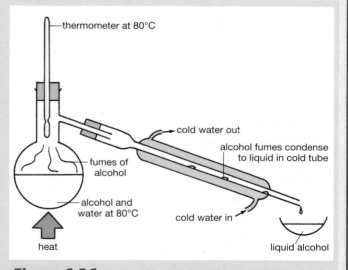

**Figure 6.36**

**Figure 6.37**

| temperature (°C) | volume of $CO_2$ released (units) | |
|---|---|---|
| | bottom-fermenting yeast | top-fermenting yeast |
| 6 | 0 | 0 |
| 8 | 5 | 0 |
| 10 | 10 | 0 |
| 12 | 11 | 1 |
| 14 | 6 | 6 |
| 16 | 1 | 10 |
| 18 | 0 | 13 |
| 20 | 0 | 12 |
| 22 | 0 | 4 |
| 24 | 0 | 0 |

**Table 6.11**

a) On the same sheet of graph paper, plot two line graphs with temperature on the horizontal (x) axis. (4)

b) In general, which type of yeast preferred
   (i)   higher temperatures?
   (ii)  lower temperatures? (1)

c) At which temperature did both yeasts work at the same rate? (1)

d) By how many times was the activity of the bottom-fermenting yeast greater than that of the top-fermenting yeast at 12°C? (1)

e) Predict the volume of carbon dioxide that would be released by the bottom-fermenting yeast at 30°C. (1)

**10** The eight experiments shown in figure 6.38 were set up to investigate factors affecting the action of an enzyme immobilised on gel pellets.

a) Which two experiments should be compared to find out if the enzyme at 20°C works better if immobilised on small pellets or large pellets? (1)

b) Which two experiments should be compared to find out if the enzyme, immobilised on large pellets, works better at 20°C or 35°C? (1)

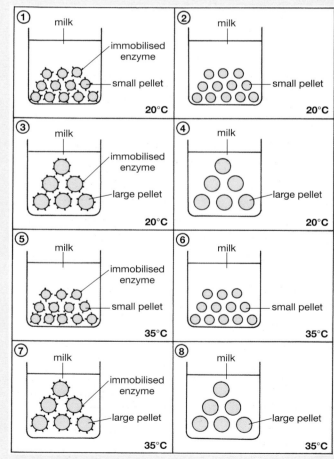

**Figure 6.38**

| additional ingredients | type of stock cube (and main ingredient) | | | | |
|---|---|---|---|---|---|
| | beef | chicken | ham | lamb | mushroom |
| chicken fat | ✗ | ✓ | ✗ | ✗ | ✗ |
| lemon juice | ✓ | ✓ | ✓ | ✓ | ✓ |
| modified starch | ✓ | ✓ | ✓ | ✓ | ✓ |
| onion | ✓ | ✓ | ✗ | ✓ | ✗ |
| pepper | ✓ | ✗ | ✗ | ✓ | ✗ |
| potato | ✗ | ✗ | ✗ | ✗ | ✓ |
| salt | ✓ | ✓ | ✓ | ✓ | ✓ |
| smoke flavouring | ✗ | ✗ | ✓ | ✗ | ✗ |
| soy sauce | ✓ | ✗ | ✗ | ✓ | ✗ |
| sugar | ✓ | ✓ | ✓ | ✓ | ✗ |
| vegetable oil | ✗ | ✗ | ✗ | ✗ | ✓ |
| yeast extract | ✓ | ✓ | ✓ | ✓ | ✗ |

key    ✓ = present
       ✗ = absent

**Table 6.12**

c) Which two experiments should be compared to find out if, using large pellets at 35°C, the enzyme must be present for the experiment to work? (1)

d) Why would a comparison of experiments 3 and 5 *not* allow a valid conclusion to be made? (1)

11 Table 6.12 shows the ingredients present in a certain brand of stock cubes used to make soups and gravies.

a) How many of the additional ingredients are found in all of the stock cube types? (1)

b) How many of the additional ingredients are found in only one type of stock cube? (1)

c) Which TWO types of stock cube have exactly the same set of additional ingredients? (2)

d)(i) How many types of stock cube do *not* contain yeast extract?

(ii) What percentage of stock cube types do contain yeast extract?

(iii) Why is yeast extract often added to stock cubes? (3)

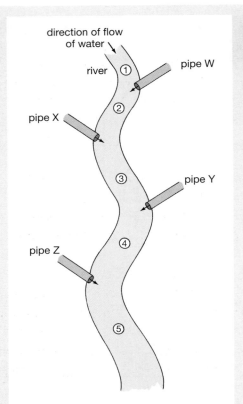

direction of flow of water

river ①  pipe W

② 

pipe X

③  pipe Y

④ 

pipe Z

⑤ 

**Figure 6.39**

| test or check | number of sample site | | | | |
|---|---|---|---|---|---|
| | 1 | 2 | 3 | 4 | 5 |
| oxygen content | high | low | medium | high | high |
| number of bacteria | low | high | medium | low | low |
| number of fish | high | low | low | medium | high |

**key**   ■ high
          ▨ medium
          □ low

**Table 6.13**

**12** Figure 6.39 shows a river with four pipes (W, X, Y and Z) pouring liquids into it. Scientists took water samples at sites 1–5. These samples were tested for oxygen content and number of bacteria. In addition the sites were checked for the number of fish present. The results are shown in table 6.13.

a) Which sample point had water containing a high number of bacteria? (1)

b) Which sample point had water with a low oxygen content? (1)

c) (i) From which pipe had wastes from yeast-based industry been released into the river?

   (ii) Explain your answer. (2)

d) Why were very few fish found at sample sites 2 and 3? (1)

e) Which TWO sites had the same oxygen content and numbers of bacteria and fish? (1)

f) Imagine methylene blue dye has been added to water samples from sites 2, 3, 4 and 5.

   (i) Which sample would stay blue for the longest time?

   (ii) Which sample would lose its blue colour fastest?

   (iii) Explain your answers to (i) and (ii). (3)

## What you should know

| | | |
|---|---|---|
| alcohol | fermentation | oxygen |
| ale | fermented | pollute |
| aseptic | flame | red hot |
| autoclave | flavour | reused |
| brewery | fungus | rise |
| carbon dioxide | gel | sugar |
| cask | immobilised | temperature |
| colour | lactic acid | time |
| dough | microscope | upgraded |
| enzyme | nutrient agar | yeast |

**Table 6.14**  *Word bank for chapter 6*

1  Yeast is a simple _____. When magnified under a _____, it is seen to be made up of single cells. Yeast can be grown on jelly-like food called _____ or in liquid culture rich in _____.

2  During laboratory work with micro-organisms (such as yeast), precautions are taken to try to create sterile (_____) conditions. For example a special container called an _____ is used to sterilise apparatus and nutrient agar by heating them to 121°C for 20 minutes. A Bunsen burner is used to heat a wire inoculating loop until it is _____ and to _____ the neck of any culture bottle in use.

3  When added to bread _____, yeast produces bubbles of _____ gas. These get caught in the dough and make it _____.

4  During beer-making, yeast is used to change sugar into _____ and carbon dioxide. This chemical reaction is called _____.

5  The alcohol content of beer varies. It is affected by factors such as the type of _____ used, the _____ at which the process is carried out and the length of fermentation _____.

6  _____-conditioned beer has its yeast removed and extra carbon dioxide added. _____-conditioned beer (real _____) does not have its yeast removed and the yeast continues to produce carbon dioxide in the cask.

7  An _____ is a substance that controls a biochemical reaction. Enzymes from bacteria can change milk sugar to simpler sugars and _____. Enzymes in yeast cells can change simple sugars to alcohol. These enzymes from microbes can change milk into a _____ milk drink.

8  Enzyme molecules and yeast cells can be _____ (trapped) in _____ pellets. After they have brought about the required chemical reaction, they can be _____.

9  Yeast is used to _____ certain foods. Pink yeast can be used to _____ the flesh of farmed salmon.

10  If wastes from yeast-based industry were released into a river, they would _____ the water. Bacterial numbers in the river water would rise and _____ content would drop. To prevent this happening and to increase profit, the waste is _____ into animal feed.

## 7 Detergent industries

## Enzymes

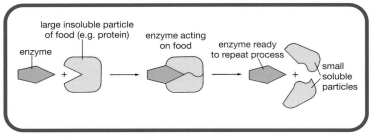

**Figure 7.1** *Action of a digestive enzyme*

**Enzymes** are substances made by all living cells. Enzymes *speed up* the rate of biochemical reactions. They normally work best at moderate temperatures (e.g. 40°C) but are destroyed at temperatures above 60°C. In the human body, digestive enzymes in the stomach and small intestine promote the breakdown (**digestion**) of large insoluble particles of food to small soluble particles (see figure 7.1).

### Activity

## Investigating the action of pepsin on protein

### Information

**Pepsin** is a digestive enzyme that is normally active in the human stomach. Egg white suspension is made of large particles of egg white **protein** that are insoluble in water and give the liquid a *cloudy* appearance.

### You need

- two labels
- two test tubes
- test tube stand
- dropping bottle of 5% cloudy egg white suspension
- dropping bottle of 1% pepsin enzyme
- dropping bottle of water
- dropping bottle of 0.1M hydrochloric acid
- two rubber stoppers
- access to water bath at 40°C

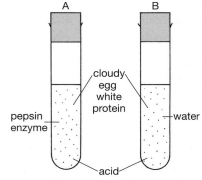

**Figure 7.2** *Investigating the action of pepsin*

### What to do (also see figure 7.2)

1 Label the test tubes A and B.
2 Add 50 drops of egg white to both tubes.
3 Add 25 drops of pepsin to tube A and 25 drops of water to tube B.
4 Add 25 drops of acid to both tubes.
5 Put the stoppers in place and invert the tubes twice to mix the contents. (Note: Tube B, the **control**, only differs from tube A by the one variable factor being investigated. All other factors have been kept the same.)

6 Place the tubes in the water bath at 40°C.

7 After 30 minutes, collect the tubes and inspect them for clearing.

8 Write a sentence to report the results of your experiment.

9 Draw a conclusion about the action of the enzyme on the insoluble particles of egg white protein.

10 If you have time, repeat the experiment and see if you get the same results. If you do, your results are even more **reliable** than before.

11 Answer the following questions:

    a) State the one variable factor by which tubes A and B differed.

    b) State two variables that were kept the same for tubes A and B.

# 'Biological' detergents

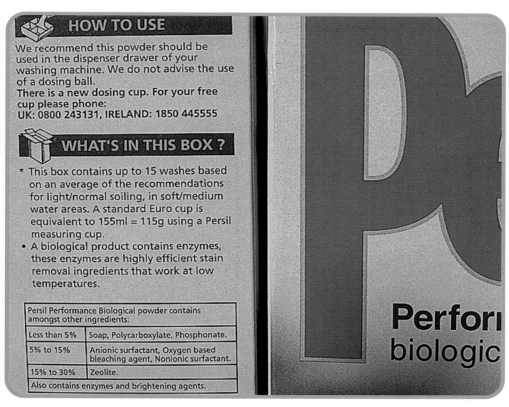

A **detergent** is a cleansing agent such as washing powder. A **'biological' detergent** is a washing powder or liquid which contains **enzymes** (see figure 7.3). These enzymes make up only about 1% of the powder yet they are so powerful that they are able to remove stains. They do this by **digesting** large particles (e.g. protein) into smaller particles which get washed away as shown in figures 7.4 and 7.5. A non-biological powder does not contain these enzymes.

**Figure 7.3** *An example of a biological detergent*

## Bacterial enzymes

The enzymes used in biological detergents are produced by **bacteria**. Bacteria are used to produce the enzymes because bacteria grow very rapidly and can be cultured in huge numbers in an **industrial fermenter** (see figure 7.6). This equipment is designed to provide the bacteria with all their growth requirements (such as food and oxygen).

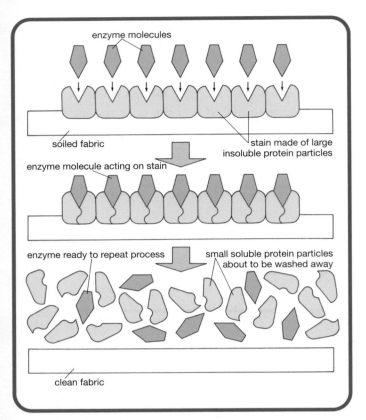

**Figure 7.4**  *Direct action of an enzyme on a stain*

As the bacteria multiply, they produce enzymes and pass them out into the surrounding culture liquid. The bacteria are then filtered off leaving a liquid rich in enzymes. The enzymes are extracted, purified and added to the washing powder or liquid to make '**biological**' products.

## Allergy

In some people, the enzymes in biological washing powders trigger **allergic reactions** such as skin rashes and eczema after prolonged contact. For this reason, the enzyme particles in biological washing powders are now **enclosed** in a harmless coating of wax (with a low melting point).

However, some people continue to show an allergic reaction to biological washing powders. It is thought that their skin is reacting to some enzyme particles that have not been completely rinsed away during the washing process and have been left behind in the fabric of the clothes. These people are advised to use a **non-biological** powder.

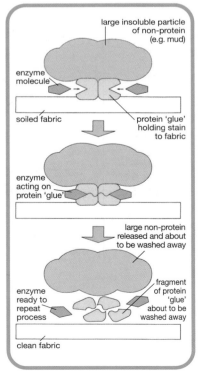

**Figure 7.5**  *Indirect action of an enzyme on a stain*

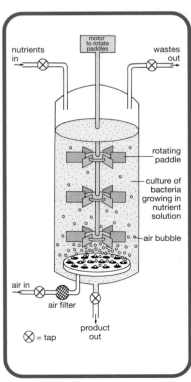

**Figure 7.6**  *An industrial fermenter*

## Activity

# Creating a home-made biological washing powder

## Information

Biological and non-biological washing powders of the same brand often differ from one another in several ways. This prevents a *fair comparison* being made between them in an experiment. To solve this problem, a home-made biological washing powder can be made by adding a **protease** (protein-digesting enzyme) to a non-biological washing powder. Then the only difference between the two powders is the *presence* or *absence* of the enzyme.

## You need

- two labels
- two dropping bottles (100 cm³)
- measuring cylinder (100 cm³ plastic)
- beaker (250 cm³ plastic)
- protease enzyme (e.g. alcalase)
- non-biological washing powder
- two weigh boats
- access to electronic balance
- stirring rod

## What to do (also see figure 7.7)

1 Label the dropping bottles 'bio w.p.' and 'non-bio w.p.' and add your initials.
2 Measure out 100 cm³ of water into the plastic beaker.
3 Weigh out 1 g of protease enzyme and 10 g of non-biological washing powder.
4 Using the stirring rod, dissolve the enzyme and powder in the water to form the solution of biological washing powder.
5 Pour this solution into the dropping bottle marked 'bio w.p.'
6 Rinse the beaker and stirring rod.
7 Measure out 100 cm³ of water into the plastic beaker.
8 Weigh out 10 g of non-biological washing powder.
9 Using the stirring rod, dissolve the powder in the water to form the solution of non-biological washing powder.
10 Pour this solution into the dropping bottle marked 'non-bio w.p.'
11 Keep these solutions for use in future activities.

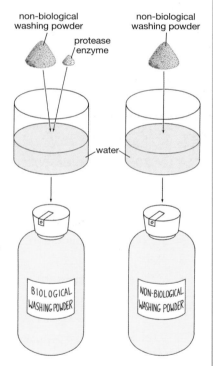

**Figure 7.7** *Creating home-made biological washing powder*

## Activity

## Testing washing powders on stained fabric

You need
- two labels
- two test tubes
- test tube stand
- dropping bottle of biological washing powder solution from previous activity
- dropping bottle of non-biological washing powder solution from previous activity
- two pieces of cloth stained with brown soy sauce (or red curry paste)
- two rubber stoppers
- access to water bath at 45°C

What to do (also see figure 7.8)

1 Label the test tubes A and B and add your initials.
2 Half-fill tube A with biological washing powder solution.
3 Half-fill tube B with non-biological washing powder solution.
4 Add a piece of stained cloth to each tube and fit the rubber stoppers.
5 Shake both tubes thoroughly and place them in the water bath at 45°C.
6 After 10–15 minutes, shake each tube again and return it to the water bath.
7 After 20–30 minutes, collect the tubes and examine the pieces of cloth for clearing of stain.
8 Describe the appearance of each piece of cloth choosing one of the following phrases:
   - stain completely cleared
   - stain mostly cleared
   - stain slightly cleared
   - stain not cleared at all
9 Draw conclusions by answering the following questions:
   a) Which of the two powders works better at 45°C?
   b) What extra substance does this powder contain?
   c) By what means does this substance bring about its effect?

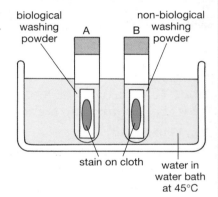

**Figure 7.8** *Testing washing powders on stained fabric*

## Activity

# Testing washing powders on milk proteins

### Information

Agar jelly is normally clear in appearance. Milk contains large particles of protein that are insoluble in water. When milk is added to agar, these particles give the agar a *cloudy* appearance and it is called **milk agar**.

### You need

- felt tip marker
- two Petri dishes of sterile milk agar
- cork borer (large, e.g. size 8)
- dropping bottle of biological washing powder solution from earlier activity
- dropping bottle of non-biological washing powder solution from earlier activity
- access to incubator at 45°C

| distance away from edge of well (mm) | appearance of agar (cloudy/clear) | | | |
|---|---|---|---|---|
| | dish A | | dish B | |
| | at start | after 24 hours | at start | after 24 hours |
| 1 | | | | |
| 10 | | | | |
| 20 | | | | |

**Table 7.1**  *Milk agar results*

### What to do

1. Prepare a copy of results table 7.1.
2. Write A on the base of the first Petri dish (near the edge) and add your initials and the date.
3. Write B on the base of the second dish and add your initials and the date.
4. Turn the dishes over and open them up.
5. Using the cork borer, cut three wells in each dish (see figure 7.9).
6. Half-fill each well in dish A with biological washing powder solution.
7. Half-fill each well in dish B with non-biological washing powder solution.
8. Put the two dishes in the incubator at 45°C.
9. After 24 hours, inspect the dishes (see figure 7.10).
10. Complete your table of results using the words 'cloudy' and 'clear'.
11. Write a sentence to state which type of washing powder was better at clearing the milk agar.
12. Answer the following questions.
    a) Agar jelly is normally clear. What gives *milk* agar its cloudy appearance?
    b) (i) What substance, present in biological powder but absent from non-biological powder, is active in dish A only?
       (ii) What effect did this substance have on cloudy milk agar's appearance after 24 hours?
       (iii) By what means did it bring about this effect?

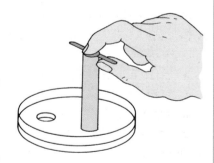

**Figure 7.9**  *Cutting a well in agar jelly*

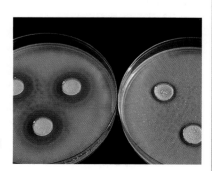

**Figure 7.10**  *Effect of protein-digesting enzyme on milk agar*

167

## Activity

# Testing washing powders on photographic film

### Information

The black chemical on black-and-white film is held onto the film by **protein 'glue'** (see figure 7.11). Similarly, stains on fabrics are often held on by protein 'glue'.

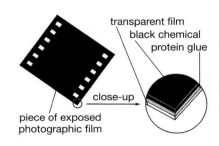

**Figure 7.11** *Role of protein glue in photographic film*

### You need

- two labels
- two test tubes
- test tube stand
- dropping bottle of biological washing powder solution from earlier activity
- dropping bottle of non-biological washing powder solution from earlier activity
- two strips of used black-and-white photographic film (10 mm wide)
- two rubber stoppers
- access to water bath at 45°C

### What to do (also see figure 7.12)

1. Label the test tubes A and B and add your initials.
2. Half-fill A with biological washing powder solution and B with non-biological washing powder solution.
3. Add a small strip of photographic film to each tube and fit the stoppers.
4. Place the tubes in the water bath at 45°C.
5. Construct a table of results.
6. After 10 minutes, gently shake the tubes and return them to the water bath.
7. After 20–30 minutes, collect the tubes and check if the film is still covered with black chemical.
8. Complete your table of results.
9. Write a sentence to say which type of washing powder was better at clearing the black chemical from the film.
10. Answer the following questions:
    a) What substance is present in biological washing powder solution but absent from non-biological washing powder solution?
    b) What effect did this substance have on the black chemical glued to the film in tube A?
    c) By what means did it bring about this effect?
    d) How could the reliability of the results be improved?

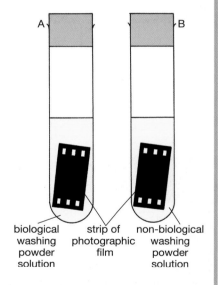

**Figure 7.12** *Testing washing powders on photographic film*

# Value and use of product

## Biological washing powder

Advantages of using biological washing powders are as follows.

- *Fuel costs are reduced.* Compared with non-biological washing powder, biological powder works better at moderate temperatures (e.g. 45°C) so less energy is needed to heat up the water for the wash.
- *Less damage is done to delicate fabrics.* Fabrics such as acrylic and wool become damaged or shrink at high temperatures such as 90°C. This does not happen at lower temperature washes. If such fabrics are badly soiled, only a biological powder will remove the stains at the lower temperature.
- *Certain 'difficult' stains can be removed.* Some stains such as grass and blood can be digested and completely removed by biological washing powder but not by non-biological powder even at high temperatures.

## Variety of enzymes

A biological detergent often contains several types of enzyme such as:

- **protein-digesting** enzyme to remove blood, grass and egg stains;
- **starch-digesting** enzyme to remove starch stains;
- **fat-digesting** enzyme to remove fat, oil and grease stains.

## Activity

## Biological enzyme assay (assessment)

### Information

**Alcalase** is the protein-digesting enzyme present in biological washing powder. In this experiment one of the three bottles (marked A, B and C) contains a concentrated solution of alcalase. One of them contains a dilute solution of alcalase and one contains water. Your job is to identify which is which by carrying out the following experiment.

### You need

- three labels
- three test tubes
- test tube stand
- dropping bottle of liquid A
- dropping bottle of liquid B
- dropping bottle of liquid C
- three strips of used black-and-white photographic film (10 mm wide)
- three rubber stoppers
- access to water bath at 45°C

# Biotechnological Industries

## What to do (also see figure 7.13)

1 Label the three test tubes A, B and C and add your initials.
2 Half-fill each tube with the appropriate liquid from a dropping bottle.
3 Add a strip of photographic film to each tube and fit a stopper.
4 Place the tubes in the water bath at 45°C.
5 Construct a table of results.
6 After 10 minutes, gently shake the three tubes and return them to the water bath.
7 After 20–30 minutes, collect the tubes and check if the film is still covered with black chemical or not.
8 Complete your table of results.
9 Identify which liquid is concentrated alcalase solution, which is dilute alcalase solution and which is water.
10 Add another column (or row) to your table to include this information.
11 Ask the teacher to assess your work.

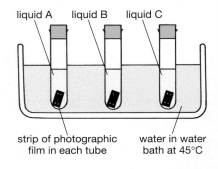

liquid A    liquid B    liquid C

strip of photographic    water in water
film in each tube        bath at 45°C

**Figure 7.13**    *Enzyme assay*

## Testing your knowledge

1 Redraw figure 7.14 and then use arrows to connect each term to its correct description. (3)

2 Briefly describe the effect that a digestive enzyme has on large insoluble particles of food. (2)

3 a) (i) What type of substance is present in biological washing powder but absent from non-biological washing powder?
   (ii) Why are bacteria used to make this type of substance? (2)
  b) Name THREE conditions needed by growing bacteria that are provided by a fermenter. (3)

4 a) Give an example of an allergic reaction that could be triggered by biological washing powder. (1)
  b) In what way are enzymes in biological washing powders now treated to make them safe to handle? (1)

5 State TWO advantages of using biological washing powder rather than non-biological washing powder for the family wash. (2)

| detergent | container in which micro-organisms are cultured in large numbers |
| enzyme | cleansing agent such as washing powder or liquid |
| fermenter | substance that speeds up the rate of a biochemical reaction |

**Figure 7.14**

# Environmental impact

**Figure 7.15** *A fossil fuel power station where electricity is generated*

## Reduced fuel consumption

As the human population continues to increase, the worldwide demand for **energy** increases. In Britain, much of the energy used as electricity in people's homes comes from **power stations** (see figure 7.15).

Many of these power stations generate energy by burning **fossil fuels** such as coal, oil or gas. This often results in the release of gases such as **carbon dioxide** ($CO_2$) and **sulphur dioxide** ($SO_2$) into the atmosphere (see figure 7.16).

### Carbon dioxide

The carbon dioxide in the air surrounding the Earth helps to keep the planet warm. However the increasing level of carbon dioxide in the atmosphere is causing the temperature of the Earth to rise. This effect is called **global warming**. It has been happening gradually for many years (see figure 7.17) and is expected to cause problems such as flooding and changes in climate in the future.

### Sulphur dioxide

If sulphur dioxide is released into the air, it can react with water and oxygen high up in the atmosphere to form acid rain clouds. When

**Figure 7.16** *'Mind if I smoke?'*

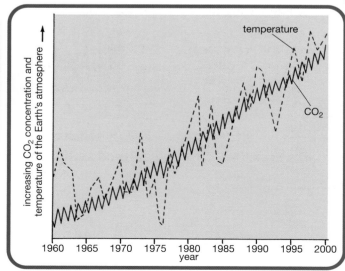

**Figure 7.17** *Effect of increasing carbon dioxide concentration on Earth's temperature*

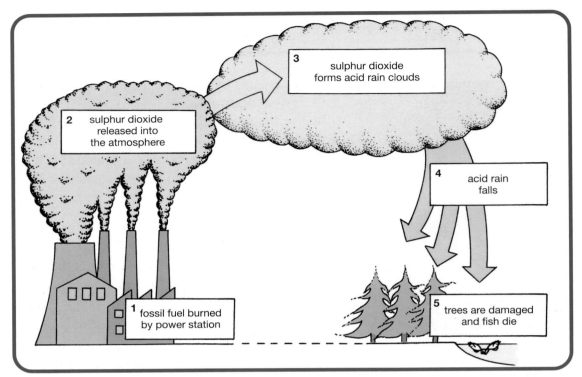

**Figure 7.18**   *Pollution by sulphur dioxide*

**Figure 7.19**   *Effect of acid rain on trees*

these fall as **acid rain**, the environment becomes **polluted** and plants and animals suffer serious damage (see figures 7.18 and 7.19).

## What can people do to help?

If more people used biological washing powder, energy could be saved because:

- biological powders work at lower temperatures so *less energy* is needed to heat the washing water;

- lower temperature programmes are normally shorter than higher temperature ones so *less energy* is needed to operate the washing machine.

Since less energy would be needed, less fossil fuel would need to be burned at the power stations and pollution would be reduced.

# Effect of detergents on wildlife

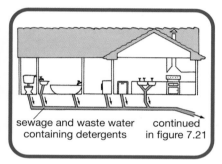

**Figure 7.20** *Detergents in household waste water*

sewage and waste water containing detergents

continued in figure 7.21

## Sewage works

**Detergents** are normally rich in chemicals called **phosphates** which help to make them highly effective cleansing agents. Detergents and phosphates in waste water pass along pipes from people's homes (see figure 7.20) to the local **sewage works**. At the sewage works (see figures 7.21 and 7.22) the waste water and sewage go to the **primary treatment** plant. Once solids have settled out, the waste liquid is passed on to the **secondary treatment** plant. It is here that useful **bacteria**, supplied with plenty of oxygen, break down the wastes to *harmless* substances. Water containing these harmless substances and phosphates is then discharged into the local river or canal.

## Foam

In the past, detergents contained chemicals that the useful bacteria at the sewage works could not break down. In fact the useful bacteria were often killed by the detergent. The untreated soapy chemicals

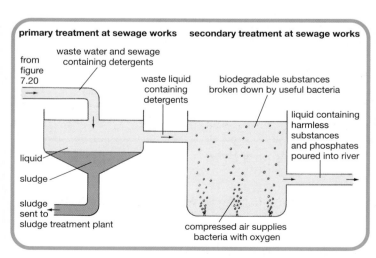

primary treatment at sewage works   secondary treatment at sewage works

from figure 7.20

waste water and sewage containing detergents

waste liquid containing detergents

biodegradable substances broken down by useful bacteria

liquid containing harmless substances and phosphates poured into river

liquid

sludge

sludge sent to sludge treatment plant

compressed air supplies bacteria with oxygen

**Figure 7.21** *Fate of waste water containing detergents*

**Figure 7.22** *An aerial view of a sewage works*

were then discharged into rivers where they became whipped up into giant clouds of **foam**. In Britain this hardly ever happens now because most modern detergents are **biodegradable**. This means that they are broken down by the useful bacteria, provided that the sewage treatment works is not overloaded with excess sewage.

## Phosphates

**Phosphates** from detergents are often released into the local river or canal after the waste water has been treated at a sewage works. Phosphates act as a form of **fertiliser** that encourages plants to grow. When river water is rich in phosphates, simple water plants called **green algae** grow so well that their population explodes forming an **algal bloom** (see figure 7.23).

Water affected in this way is so green and cloudy that it looks like pea soup. Normally the number of algal cells in a river is kept in check by water fleas eating them. But when an algal bloom forms, many algal cells are left uneaten. When they die, their dead remains acts as a food source for bacteria in the river water. These bacteria increase rapidly in number and use up the river's **oxygen** supply. River animals such as fish that depend on this oxygen suffer and may die. This series of events is summarised in figure 7.24. It shows how excessive use of detergents rich in phosphate has a poisonous (**toxic**) effect on wildlife since it leads indirectly to the death of fish (also see figure 7.25).

**Figure 7.23**   *An algal bloom*

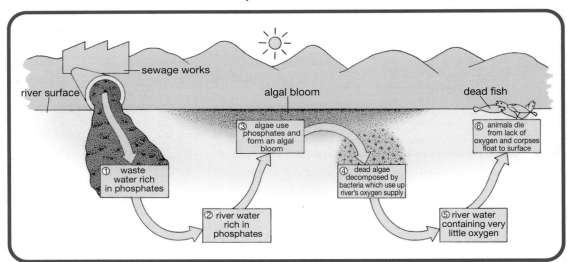

**Figure 7.24**   *The toxic effect of phosphates on a river*

**Figure 7.25** *'I'm glad you won't be needing the kiss of life!'*

## Reducing the impact of phosphate on the environment

The phosphates entering rivers, canals and lochs come from many sources and vary from one region to another. The pie chart in figure 7.26 shows the data from a typical region.

### Replacing phosphate in detergents

Some countries have tried *banning* phosphates in detergents and insisting that people use products containing a *substitute chemical*. However the phosphate-free detergents are more expensive. In addition, so much phosphate enters the rivers from other sources (see figure 7.26) that hardly any improvement to the environment has resulted so far.

### Removal of phosphates at the sewage works

In the past, sewage works often discharged waste water rich in phosphates into the local river (see figure 7.21). However many sewage treatment plants now remove about 90% of the phosphates from the water by **tertiary treatment** (see figure 7.27).

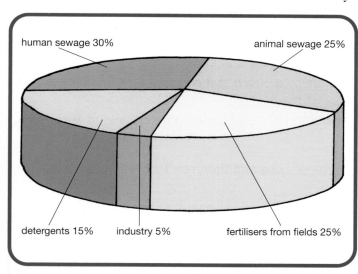

**Figure 7.26** *Sources of phosphates in river water*

One of the advantages of this solution is that the problem of phosphates in human sewage and detergent can be tackled at the same time. However the equipment is expensive to install and maintain. In addition, it can only make a real difference if similar equipment is used to remove phosphates from animal sewage.

### Use of minimum

Many people use far too much washing powder and detergent when doing the family wash. Use of the *minimum* (as stated by the manufacturer on the pack) would help to reduce the impact of detergents on the environment.

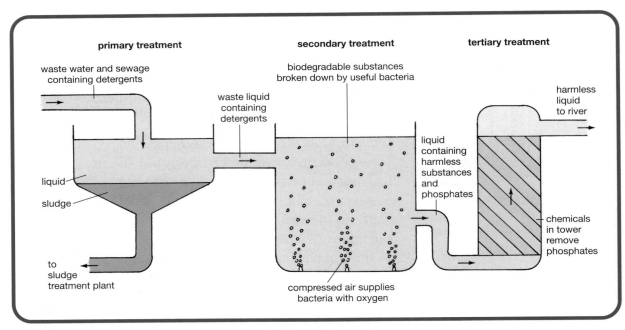

**Figure 7.27** *Fate of waste water containing detergents*

## Testing your knowledge

1 Redraw figure 7.28 and then use arrows to connect each term with its description. (3)

2 a) What TWO gases may be released into the atmosphere by a power station that burns fossil fuels? (2)
   b) State a form of pollution that can be caused by one of these gases. (1)
   c) Why might increased use of biological detergents in place of non-biological detergents help to reduce this problem? (2)

3 Decide whether each of the following statements is true or false and then use T or F to indicate your choice. If a statement is false, give the word that should have been used in place of the one in **bold** print. (5)
   a) Detergents contain chemicals that can promote the growth of **algae** in a river.
   b) **Phosphates** act on dead algal cells and decompose them.
   c) An increase in number of bacteria in a river leads to an **increase** in oxygen content of the water.
   d) Waste water rich in detergent can be **toxic** to wildlife.
   e) Phosphates in waste water are often removed by tertiary treatment at a **fermenter** works.

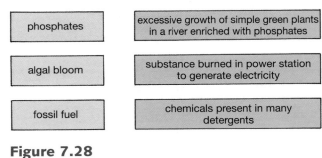

**Figure 7.28**

## Applying your knowledge

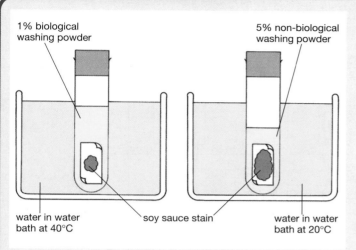

1% biological washing powder

5% non-biological washing powder

water in water bath at 40°C

soy sauce stain

water in water bath at 20°C

**Figure 7.29**

1 The experiment shown in figure 7.29 was set up to compare the effect of a biological and a non-biological washing powder on soy sauce stains at 40°C. State FOUR ways in which the experiment must be altered to make it a valid test. (4)

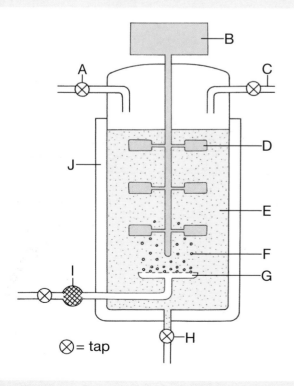

⊗ = tap

**Figure 7.30**

2 Figure 7.30 shows a simple diagram of an industrial fermenter being used to grow bacteria which produce a useful enzyme.
   a) Which lettered structure indicates each of the following?
      (i) an air bubble;
      (ii) a filter to clean air before it enters the fermenter;
      (iii) a rotating paddle;
      (iv) the culture of bacteria;
      (v) the tap that would be opened at the end of the process to release the enzyme and the bacteria. (5)
   b) The boxes in figure 7.31 show four stages involved in producing an enzyme for use in a biological washing powder. Draw a flow diagram of the four stages in the correct order. (2)

3 The four experiments shown in figure 7.32 on page 178 were set up to investigate the action of a new biological soap powder called BIOX.

bacterial cells are poured off and separated from the culture liquid by filtration

bacteria are put into the fermenter and given ideal conditions for growth

the enzyme is extracted from the culture liquid and added to the washing powder

bacterial cells are allowed to multiply and release the enzyme into the surrounding culture liquid

**Figure 7.31**

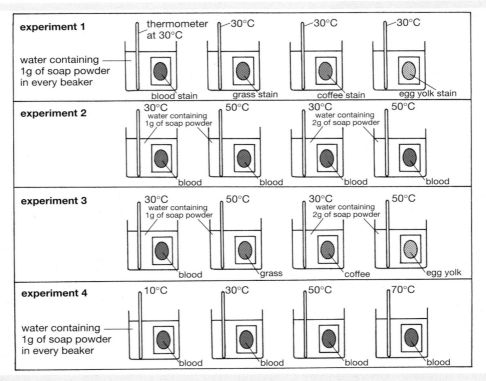

**Figure 7.32**

Study them carefully and answer the following questions.

a) Which experiment is testing the effect of four different temperatures on the action of BIOX? (1)

b) In which experiment is the type of stain the one variable factor being investigated? (1)

c) Which experiment really consists of two experiments being done at the same time? (1)

d) Which experiment is *not* valid? Explain why. (2)

**4** Table 7.2 shows the results of an experiment set up to test the action of two detergents

| temperature (°C) | percentage of stain removed | |
| --- | --- | --- |
| | detergent A | detergent B |
| 10 | 15 | 15 |
| 20 | 30 | 20 |
| 30 | 60 | 24 |
| 40 | 100 | 30 |
| 50 | 70 | 40 |
| 60 | 62 | 60 |
| 70 | 78 | 80 |
| 80 | 98 | 96 |
| 90 | 100 | 100 |

**Table 7.2**

(A and B) on mud stains at different temperatures.

a) On the same sheet of graph paper, draw two line graphs of these results. (Put temperature on the horizontal x-axis.) (5)

b) (i) At which temperature did detergent A remove 60% of the stain?

(ii) What percentage of the stain was removed by detergent A at 50°C? (2)

c) What were the best two temperatures for detergent A to work? (1)

d) (i) At which temperature did detergent B remove 60% of the stain?

(ii) What percentage of the stain was removed by detergent B at 70°C? (2)

e) What was the best temperature for detergent B to work? (1)

f) One of the detergents was non-biological washing powder and the other was the same non-biological powder with a protein-digesting enzyme added. Identify the two powders and explain your answer. (3)

5 Figure 7.33 shows eight experiments. They were set up to investigate the effect of two temperatures and two pH values on a biological and a non-biological detergent in contact with milk agar.

a) Which TWO Petri dishes should be used to find out:

(i) the effect of biological detergent compared with non-biological detergent at 20°C and pH 7?

(ii) the effect of biological detergent compared with non-biological detergent at 40°C and pH 4? (2)

b) Which TWO Petri dishes should be compared to find out:

(i) the effect of pH on biological detergent at 40°C?

(ii) the effect of temperature on biological detergent at pH 7?

(iii) the effect of temperature on non-biological detergent at pH 4? (3)

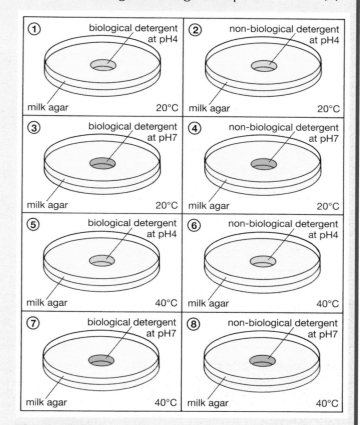

**Figure 7.33**

| percentage of black chemical stuck onto film | | concentration of washing powder (%) | | |
| | | 1 | 2 | 3 |
|---|---|---|---|---|
| percentage of black chemical stuck onto film | at start | 100 | 100 | 100 |
| | after 1 hour | 50 | 20 | 10 |

**Table 7.3**

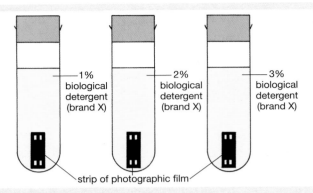

**Figure 7.34**

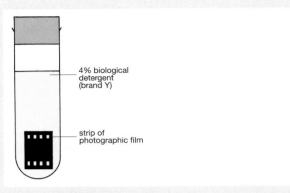

**Figure 7.35**

**6** A type of protein glue is used to attach the black chemical to photographic film. This glue can be digested by a biological detergent. Figure 7.34 shows an experiment set up to investigate the effect of the detergent's concentration on its activity. Table 7.3 on page 179 shows the results after one hour.

a) What type of chemical in biological detergent digests the protein glue? (1)

b) What was the ONE variable factor investigated in this experiment? (1)

c) Give TWO variable factors that were kept the same when this experiment was set up. (2)

d) What conclusion can be drawn from the results? (1)

e) Suggest how the reliability of the results could be improved. (1)

f) The scientist decided to try a concentration of 4% detergent. She set up the test tube shown in figure 7.35. Give TWO reasons why this set-up would not give valid results. (2)

**7** Read the passage on page 181 and answer the following questions.

a) When microbiologists want to find out if a microbe should be grown on a large scale to produce a useful enzyme, they first carry out experiments on it. Look at figure 7.36 and suggest THREE experiments that they would carry out. (3)

b) (i) Copy and complete table 7.4 using the information in the passage.

(ii) One of the earliest biological detergents was designed to remove blood protein stains from butchers' aprons. Which enzyme in the table would do this job? (7)

c) How many grams of enzyme will be present in a packet of biological powder that weighs (i) 800 g? (ii) 1.1 kg? (2)

d) (i) By what means was denim first given its stonewashed appearance?

(ii) What was the disadvantage of this method?

(iii) Which enzyme can be used to give the stonewashed effect?

(iv) Give TWO advantages of this more recent method. (5)

| enzyme | trade name | source | use |
|---|---|---|---|
| amylase | termanyl | bacterium | removes starch stains |
| lipase | | fungus | removes grease stains |
| | alcalase | | |
| cellulase | celluclast | | |

**Table 7.4**

## STUBBORN STAIN SHIFTERS

Enzyme technology is a multi-million pound industry in many developed countries. In recent years, scientists have found ways of extracting many useful enzymes from microbes. The microbiologist's first job is to select a suitable microbe. Some of the factors to be considered are shown in figure 7.36.

The detergent industry is the largest single market for these enzymes. It accounts for 30% of total sales worldwide. This industry makes use of several different types of enzyme. Proteases (such as **alcalase**) are added to digest protein stains. Amylases (such as **termanyl**) are used to shift starch stains. Both of these types of enzyme are made by bacteria. Lipases (such as **lipolase**) are made by fungi and remove fat and grease stains. These enzymes are so powerful that they only make up 1% of a biological soap powder's ingredients.

The blue dye that gives denim jeans their colour is held within cellulose fibres in the cloth. Hardwearing denim clothing was originally designed for American cowboys but it has now become popular leisurewear with faded 'stonewashed' jeans often enjoying high fashion status. Originally the stonewashed appearance was created by rotating the cloth in a drum of stones. The stones broke down some of the fibres and released the dye. However, this often caused serious damage to the garment.

Recently scientists have found ways of extracting enzymes called cellulases (such as **celluclast**) from fungi. When areas of a denim garment are soaked in cellulase, the enzyme breaks down the cellulose fibres at the surface and releases the dye. This causes minimum damage and can be used to make some parts of the garment go faded while leaving the rest unchanged (see figure 7.37).

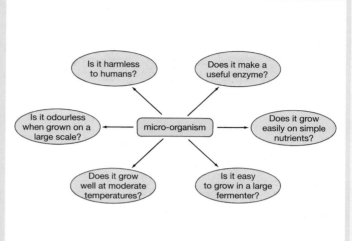

**Figure 7.36**

**Figure 7.37**

8 Figure 7.38 shows the chain of events that occurs when river water is polluted with excess phosphate from detergents.
Match boxes A–E with the following numbered statements:
1) Many algal cells die.
2) Water flea numbers do not increase quickly enough to keep the algae in check.
3) Waste water rich in phosphate is added to the river.
4) Bacteria decompose dead algae and use up the river's oxygen supply.
5) Algae use phosphate and a population explosion occurs. (5)

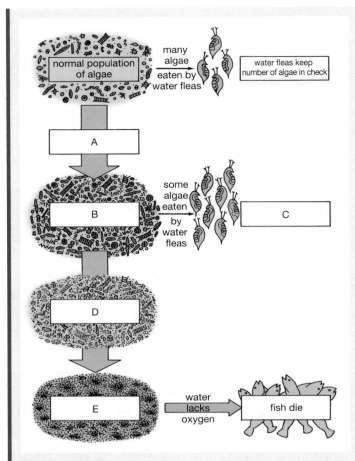

**Figure 7.38**

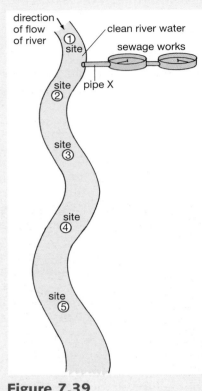

**Figure 7.39**

| source of phosphate | percentage |
|---|---|
| 1  human sewage | 32 |
| 2  farm animal sewage | 25 |
| 3  fertiliser washed off fields by rain | 20 |
| 4  industry | 7 |
| 5  detergents | 16 |

**Table 7.5**

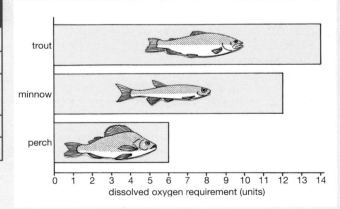

**Figure 7.40**

**9** Table 7.5 shows sources of phosphate found in a British river.

a) Draw a bar chart of these data. (3)

b) By how many *per cent* is the phosphate from farm animal sewage greater than that from detergents? (1)

c) By how many times is the percentage of phosphate from human sewage greater than that from detergents? (1)

**10** Treated liquid waste from the sewage works is poured into the river in figure 7.39 at pipe X. In 1982 this liquid was rich in phosphate from detergents.

a) An algal bloom was found at site 2. Explain how this developed. (1)

b) Rotting algae and dead fish were found at site 3. Suggest what caused the death of the fish. (2)

c) All three types of fish shown in figure 7.40 were found at site 1. However no live fish were found downstream from pipe X until site 5. There, one of the three types had reappeared. Suggest which one and explain your answer. (2)

d) In 2002 the survey was repeated and all three types of fish were found at all five sample sites and phosphate levels had dropped. Suggest the way in which the sewage works had been altered. (1)

## What you should know

| | | |
|---|---|---|
| algae | electricity | number |
| allergic | enclosed | oxygen |
| bacteria | energy | phosphates |
| banned | enzyme | pollution |
| biological | fermenters | power |
| cells | fertiliser | river |
| damage | fish | sewage |
| detergent | fossil | sulphur |
| digest | high | toxic |
| eczema | moderate | waste |

**Table 7.6**  *Word bank for chapter 7*

1 An _____ is a substance that speeds up the rate of a biochemical reaction. All living _____ make enzymes.

2 Large quantities of useful enzymes can be produced by culturing the _____ that make them in industrial _____.

3 A _____ is a cleansing agent such as washing powder. A _____ detergent contains enzymes able to _____ stains; a non-biological detergent does not contain enzymes.

4 Some people are _____ to the enzyme in biological washing powders and develop skin rashes or _____. The enzymes are now _____ in a harmless coating of wax to allow them to be handled safely.

5 Whereas non-biological washing powder only works well at _____ temperatures (e.g. 90°C), biological washing powder works well at _____ temperatures (e.g. 40°C) because the enzymes in it digest stains well at these lower temperatures.

6 A lower temperature wash saves _____ and causes less _____ to delicate fabrics.

7 Since less energy is needed for a lower temperature wash, the demand made on the _____ stations that generate _____ by burning _____ fuel is reduced. Decreased combustion of fossil fuel cuts down the quantities of harmful gases such as _____ dioxide released into the atmosphere. As a result _____ of the environment is reduced.

8 Detergents contain chemicals called _____. If waste water containing phosphates is released into a river or loch, the phosphates may act as _____ and encourage the growth of huge populations of _____.

9 When the algae die, they are decomposed by bacteria which increase in _____ and use up the river's _____ supply. Lack of oxygen in the water results in the death of river animals such as _____. The detergent is therefore said to be _____ (poisonous) to wildlife.

10 In some countries, phosphates are _____ from use in detergents and replaced by other chemicals. A more effective method of controlling the phosphate content of _____ water is to remove them at a _____ works before the water is released into the local _____.

## 8  Pharmaceutical industries

### Antibiotics

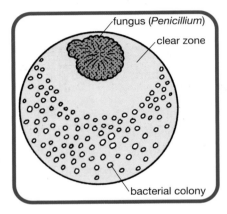

**Figure 8.1** *Fleming's famous dish*

### Discovery of penicillin

In 1928, Alexander Fleming, a Scots doctor working in a London hospital, set up an experiment to grow colonies of bacteria (called *Staphylococcus*) on nutrient agar in Petri dishes. After incubating the dishes for two days, he looked at them expecting to see the agar in each covered with colonies of bacteria. However one of the dishes was different from the others. A stray fungal spore had, by chance, entered it and grown into a healthy colony of a green fungal mould called *Penicillium* (see figure 8.1).

### First antibiotic

When Fleming looked at the 'spoiled' dish, he noticed something unusual that led him to make an important discovery. The area around the fungal colony, instead of being cloudy with bacteria, was *clear*. He therefore concluded that a chemical made by the fungus was *preventing the growth* of (or perhaps even killing) the bacteria. This chemical was later extracted and shown to work against several types of bacteria. It was the first **antibiotic** to be discovered and Fleming called it **penicillin**.

Other scientists carried out further investigations and began to develop better and better methods of extracting penicillin from the fungus and purifying it. It was not until 1941 that they had enough to try it out on a human case. The first man to be treated with penicillin was a policeman. He was so ill with blood-poisoning (septicaemia) that doctors did not expect him to live for more than a few hours. He was injected with penicillin and after 24 hours, instead of being dead, showed a *marked improvement* in his condition. He was given more penicillin and continued to make progress. However after ten days, the scientists ran out of the antibiotic and the man died.

Sad though it was, this case showed that penicillin destroys certain bacteria *without harming the human body*. Work continued and eventually scientists in Britain and USA developed methods of producing penicillin on a massive scale. Penicillin is now used to cure diseases such as pneumonia and diphtheria. Fleming was knighted during his lifetime in honour of his discovery of the first antibiotic.

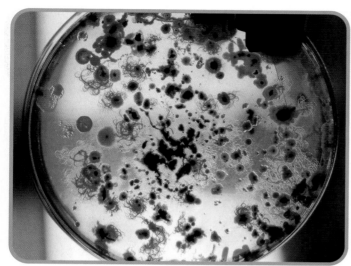

**Figure 8.2** *Soil microbes*

## Other antibiotics

Following Fleming's work, many other antibiotics have now been discovered. Each antibiotic is a naturally occurring substance produced by one type of micro-organism (e.g. a fungus) that is passed out into the surrounding environment where it prevents the growth (or even kills) other types of micro-organisms (e.g. some types of bacteria).

Fresh soil contains many types of micro-organism such as fungi and bacteria. These can be cultured on nutrient agar (see figure 8.2). Under natural conditions in the ground, some soil microbes produce antibiotics to kill their rivals and reduce the competition for food and space.

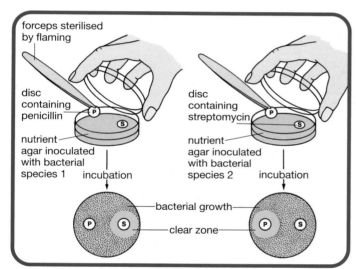

**Figure 8.3** *Action of antibiotics*

## Sensitive or resistant?

If a micro-organism's growth is prevented by an antibiotic, the microbe is said to be **sensitive** to the antibiotic. If the antibiotic has no effect, the microbe is said to be **resistant**. From the experiment shown in figure 8.3, it can be concluded that bacterial species 1 is sensitive to streptomycin and resistant to penicillin whereas bacterial species 2 is sensitive to penicillin and resistant to streptomycin.

There is no one antibiotic that works against all species of bacteria. Different antibiotics are

| antibiotic | disease caused by bacteria | | | |
|---|---|---|---|---|
| | pneumonia | tuberculosis | typhoid | diphtheria |
| penicillin | +++ | – | – | +++ |
| streptomycin | – | +++ | – | – |
| tetracycline | +++ | – | + | ++ |
| chloramphenicol | +++ | – | +++ | ++ |

key  +++  = very effective
++   = effective
+    = slightly effective
–    = no effect

**Table 8.1** *Effectiveness of different antibiotics*

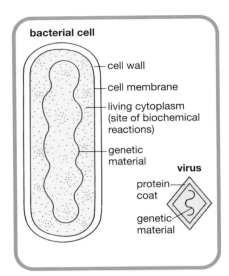

**Figure 8.4** *Structure of bacterial cell and virus*

effective against different bacteria. Also antibiotics vary in *how* effective they are, as shown in table 8.1.

## How antibiotics work

Figure 8.4 shows a simple diagram of a **bacterial cell** and a **virus**. Different antibiotics employ different methods to destroy bacterial cells as follows:

- some destroy bacterial cell walls;
- some burst the cell membrane;
- some interfere with biochemical reactions that take place in the cell's cytoplasm.

A **virus** is a tiny particle that can only multiply and cause disease once it is inside a living cell. However the virus itself is not a living cell. It does not have a cell wall, a cell membrane or living cytoplasm and it is *not affected by antibiotics*. It is for this reason that antibiotics do not work against viral infections such as the common cold and influenza.

## Activity

### Investigating the effect of antibiotic multidiscs on bacteria

#### Information

A **multidisc** (see figures 8.5 and 8.6) is a disc (or ring) of sterile filter paper. The tip of each of its arms contains a different **antibiotic**. It allows the microbiologist to test the sensitivity of a bacterium to several different antibiotics at one time.

#### You need

- felt tip marker
- Petri dish of nutrient agar inoculated with bacterial species 1
- Petri dish of nutrient agar inoculated with bacterial species 2
- Bunsen burner
- pair of metal forceps
- two antibiotic multidiscs (in a plastic bag)
- sellotape and scissors
- access to incubator at 30°C

#### What to do (also see figure 8.7)

1 Turn over the first dish marked 'bacterial species 1' and write your initials and the date on the base near the edge.
2 Repeat step 1 for the second dish marked 'bacterial species 2'.
3 Light a Bunsen burner to create an updraught to keep airborne spores away from your experiment.

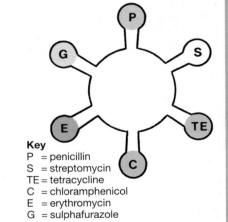

**Key**
P = penicillin
S = streptomycin
TE = tetracycline
C = chloramphenicol
E = erythromycin
G = sulphafurazole

**Figure 8.5** *A multidisc*

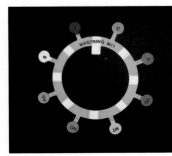

**Figure 8.6** *A ring multidisc*

4 Heat the tips of the forceps for a few seconds as shown in the diagram.

5 Allow the forceps to cool for a few seconds.

6 Use the forceps to collect an antibiotic multidisc from the plastic bag.

7 Working near the Bunsen, open dish 1 and hold the lid as a shield above the base.

8 Place the multidisc in the centre of the dish with all the antibiotics in contact with the agar and bacteria, and quickly close the dish.

9 Repeat steps 4–8 for dish 2.

10 Repeat step 4.

11 Tape the dishes and place them in the incubator at 30°C.

12 Make up a results table (like the one shown in table 8.2 with a list of letters to represent the antibiotics on the arms of your multidiscs).

13 After 36 hours inspect the dishes and complete your results table.

14 Draw conclusions by answering the following questions:
   a) To how many antibiotics was bacterial species 1 sensitive?
   b) To how many antibiotics was bacterial species 1 resistant?
   c) To how many antibiotics was bacterial species 2 sensitive?
   d) To how many antibiotics was bacterial species 2 resistant?

15 State how the results could be made more reliable.

| antibiotic | appearance of area around antibiotic disc after 36 hours (cloudy/clear) | |
| --- | --- | --- |
| | bacterial species 1 | bacterial species 2 |
| P | | |
| S | | |
| TE | | |
| C | | |
| E | | |
| G | | |

**Table 8.2** *Multidisc results*

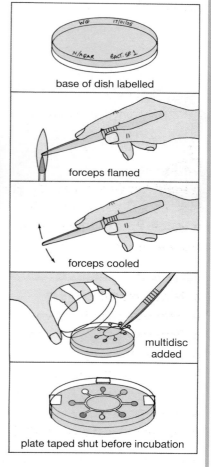

base of dish labelled

forceps flamed

forceps cooled

multidisc added

plate taped shut before incubation

**Figure 8.7** *Setting up the multidisc experiment*

## Medical use of an antibiotic multidisc

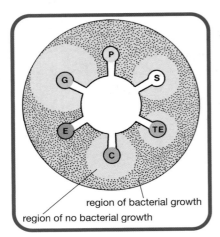

**Figure 8.8**  *Use of a multidisc*

**Figure 8.9**  *A ring multidisc in action*

When a patient is suffering from an unknown bacterial infection, it is necessary to quickly identify one or more antibiotics that will be effective against the germs. A sample of the disease-causing microbe is spread onto nutrient agar in a Petri dish and an **antibiotic multidisc** added. Several *repeat* dishes are also set up to make sure that the results are *reliable*.

Figure 8.8 shows a possible set of results after 36 hours. The bacterium is sensitive to any antibiotic that has *a clear zone* around it. In this case, therefore, antibiotics S, TE, C and G could all be considered for use in treating the infection. Figure 8.9 shows a multidisc applied to a different bacterium. In this case antibiotics S, T, C, E and FC could all be considered for use.

### Range of antibiotics

It is important that doctors have a choice of antibiotics available to treat bacterial infections for the following reasons:

- If the person is **allergic** to one antibiotic, then another one can be used.
- If the bacteria that caused the infection are **resistant** to one antibiotic (or later become resistant to that antibiotic), another one can be used.

## Antifungals

Some infections of the human body are caused by fungi. These can be spread from person to person by tiny particles of fungus called **spores**. An **antifungal** is a chemical which slows down or stops the growth of fungal infections. Many products containing antifungal chemicals are available to buy at any chemist shop. A few are shown in figure 8.10. Some examples of common fungal infections that can be successfully treated using antifungal chemicals are described below.

### Athlete's foot

The fungus that causes **athlete's foot** likes to grow on areas of the skin that are warm and moist but get little fresh air. One of its favourite

**Figure 8.10** *Antifungal treatments*

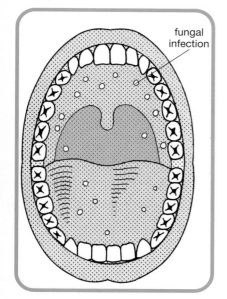

itchy fungal
infection

**Figure 8.11** *Athlete's foot*

fungal
infection

**Figure 8.12** *Oral thrush*

places is between the toes where it invades the skin, multiplies and forms an *itchy rash* (see figure 8.11).

About 1 in 7 people in Britain have athlete's foot at any one time. It is easy to catch it by walking barefoot in a place used by many people, such as a communal dressing room or shower area. A few flakes of skin from an infected person are enough to pass the infection on to many others. Athlete's foot is easily treated using **antifungal** cream and powder.

## Oral thrush

**Oral thrush** is a fungal infection of the mouth. It takes the form of white or cream-coloured spots in the mouth (see figure 8.12). It is common amongst:

- very young babies;
- people with ill-fitting dentures;
- people on antibiotics;
- people undergoing chemotherapy;
- drug users.

Oral thrush is successfully treated using **antifungal** medicines such as mouth washes and pastilles to suck. In very serious cases, the person may need a course of injections of antifungal medicine.

## Ringworm

**Ringworm** is caused by a fungus similar to the one that causes athlete's foot. It is not caused by a worm! The infection begins as a small patch of itchy skin on, for example, the inside of the thigh. If the infection is left untreated, the outer edge of the affected area spreads outwards forming ring shapes. If caught in its early stages, ringworm is easily cleared by applying **antifungal** cream. In severe cases, the person may also need to take antifungal medicine orally.

## Activity

### Investigating the effect of an antifungal on the growth of yeast

**You need**
- Petri dish containing glucose yeast agar (glucose, bactopeptone and yeast extract in agar at pH 5.8)
- felt tip marker
- Bunsen burner
- McCartney bottle of 'brown' yeast culture in liquid medium (ingredients as above without agar)

OR

- McCartney bottle of 'pink' yeast culture in liquid medium (ingredients as above without agar)
- paper towels
- access to water bath at 40°C (where McCartney bottles are kept when not in use)
- sterile syringe (1 cm$^3$)
- discard jar containing disinfectant (e.g. virkon)
- sterile spreader
- sample of antifungal treatment
- sellotape and scissors
- access to incubator (30°C for 'brown' yeast; 20°C for 'pink' yeast)

**What to do (also see figure 8.13)**
1. Turn over the dish of glucose yeast agar and write your initials and the date on the base near the edge.
2. Light the Bunsen burner to give an upward airflow near to where you are working.
3. Collect a bottle of yeast culture from the water bath and dry it on a paper towel.
4. Shake the culture gently and then unscrew the bottle's cap and draw up 0.3 cm$^3$ of yeast suspension into the syringe.
5. Using the lid as a shield, add the yeast suspension to the agar in the dish and close it.
6. Place the syringe in the discard jar and collect a sterile spreader.
7. With the lid acting as a shield again, use the spreader to spread the yeast cells evenly over the surface of the agar.
8. Place the spreader in the discard jar.
9. Add a squirt of antifungal cream to the centre of the agar.

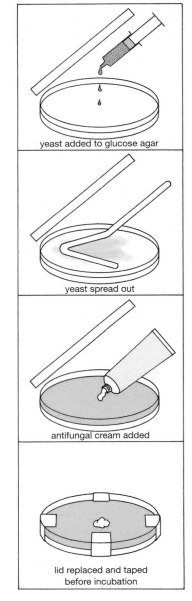

yeast added to glucose agar

yeast spread out

antifungal cream added

lid replaced and taped before incubation

**Figure 8.13** *Setting up the antifungal experiment*

10 Seal the dish using sellotape.
11 Place the dish in the incubator.
12 After 3–4 days, collect the dish and look for:
   a) colonies of healthy yeast cells;
   b) signs of antifungal activity (also see figure 8.14).
13 Draw a conclusion from your experiment about the effect of antifungal cream on the growth of yeast.
14 Describe the control experiment that should have been included in this experiment.

**Figure 8.14**   *An antifungal chemical in action*

## Testing your knowledge

1 a) Who discovered the first antibiotic? (1)
  b) When this scientist looked at one of his plates of bacteria, he noticed something unusual.
     (i) What was it? (ii) What conclusion did he come to? (2)
  c) What name was given to the first antibiotic? Suggest why. (2)
  d) Why do many soil microbes living in their natural environment make antibiotics?  (1)

2 Copy and complete figure 8.15 using arrows to connect each term to its correct meaning. (3)

3 a) Give TWO possible ways in which antibiotics work on bacterial cells and prevent their growth. (2)
  b) Do antibiotics work on viruses? Explain your answer. (2)

4 a) Name TWO fungal infections that can infect the human body. (2)
  b) What name is given to chemicals used to treat such infections? (1)

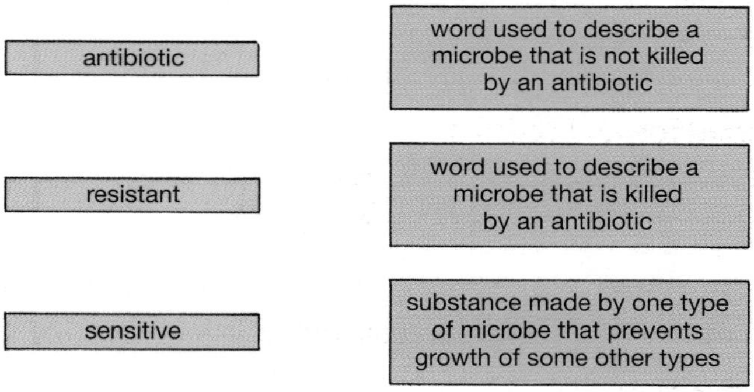

| antibiotic | | word used to describe a microbe that is not killed by an antibiotic |

| resistant | | word used to describe a microbe that is killed by an antibiotic |

| sensitive | | substance made by one type of microbe that prevents growth of some other types |

**Figure 8.15**

## Modern production methods

### Genetic engineering

Every living thing is made up of one or more tiny units called **cells**. Each cell contains **genetic material** made up of **genes**. Genes control the living thing's features and characteristics. A microbe, for example, can only make an antibiotic if it has inherited the genes that control production of that antibiotic.

In recent years, scientists have developed a technology that allows them to transfer genes from one living thing to another. This is called **genetic engineering**. The living thing that is given the 'foreign' gene is said to be **reprogrammed**. Genetic engineering allows a scientist to create micro-organisms able to make a desired product that they could not make before.

### Insulin

**Insulin** is a type of chemical called a **hormone**. It is made by the human body to control the level of sugar in the blood. People who are not able to make enough insulin are called **diabetics**. They must be given insulin or they will suffer diabetes (and perhaps die).

Figure 8.16 shows an example of genetic engineering. It results in the formation of a new type of bacterium containing the **human insulin gene**. When this type of bacterium is mass-produced, the bacteria make large quantities of **insulin** which can then be used to treat diabetics.

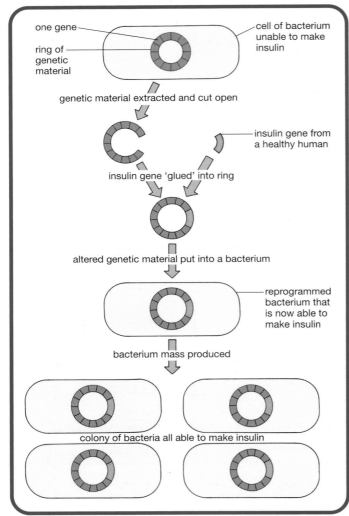

one gene
ring of genetic material
cell of bacterium unable to make insulin

genetic material extracted and cut open

insulin gene from a healthy human

insulin gene 'glued' into ring

altered genetic material put into a bacterium

reprogrammed bacterium that is now able to make insulin

bacterium mass produced

colony of bacteria all able to make insulin

**Figure 8.16**  *Genetic engineering*

### Hepatitis B vaccine

**Hepatitis B** is a serious disease of the human liver. It is caused by a virus. A gene from this virus has been inserted into yeast cells. When these yeast cells are mass-produced, they make a viral protein. When people are given an injection of this viral protein, it stimulates their immune system and protects them against the disease. A chemical that has this type of effect is called a **vaccine**.

Genetically-engineered bacteria and yeast cells are used to make many different pharmaceutical products. They are ideal for this job because they will grow rapidly in a fermenter (see below) and produce large quantities of the required substance.

## Computer-controlled technology

Biotechnological industries grow micro-organisms on a vast scale in order to produce huge quantities of a useful product such as an antibiotic. Such commercial production of antibiotics (and other substances such as enzymes and vaccines) is made possible by the use of enormous containers called **industrial fermenters** which are able to hold *thousands of litres* of nutrient liquid.

Figure 8.17 shows a simplified version of an industrial fermenter being used to grow a fungus. The system is **automated**. This means that it is controlled automatically by **computers**. **Sensors** in contact with the nutrient solution keep a check on (*monitor*) the various

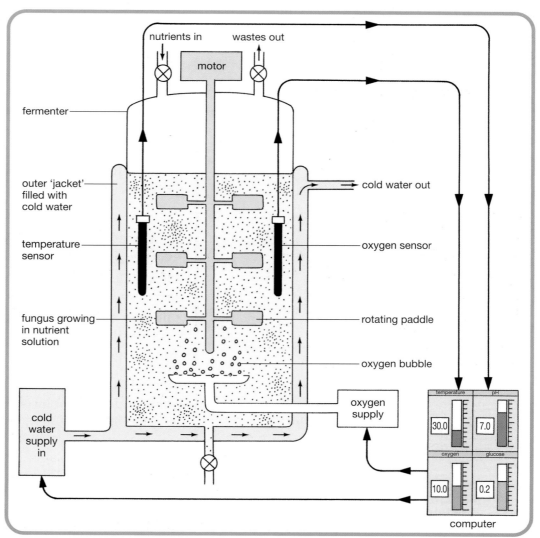

**Figure 8.17** *Industrial fermenter*

conditions that are needed for the best growth of the fungus. Some examples of factors that affect fungal growth are:

- temperature;
- oxygen concentration;
- pH;
- glucose concentration.

If any of these factors varies from the correct setting, the sensor picks up the change and sends information to the computer. The computer responds by communicating with the source of supply of the essential factor. The supply is adjusted until the required level is restored.

### Imaginary example

Let us imagine that the fungus being cultured in the fermenter in figure 8.17 grows best when given:

- 30°C;
- 10% oxygen;
- pH 7;
- 0.2 molar glucose solution.

If the temperature rises above 30°C then the **temperature sensor** picks up this information and sends it to the computer. The computer then sends out a message which causes an increase in the rate of flow of the cold water cooling system. This continues until the temperature drops back down to 30°C.

Let us imagine that the growing fungus is using up oxygen rapidly and the concentration has dropped below 10%. This information is picked up by the **oxygen sensor** and sent to the computer. The computer responds by sending out a message that increases the supply of oxygen to the fermenter until the level returns to 10%.

Although not shown in figure 8.17, further sensors for **pH** and **nutrient levels** would be monitoring these factors allowing them to be adjusted as required.

### Monitoring the product

Given ideal growing conditions, the fungus grows rapidly and produces the product which it releases into the surrounding liquid. Computer-controlled technology also *monitors* this release and brings

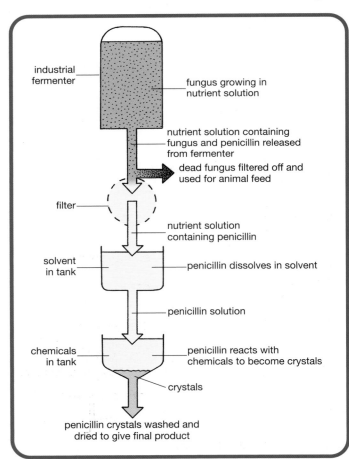

industrial fermenter

fungus growing in nutrient solution

nutrient solution containing fungus and penicillin released from fermenter

dead fungus filtered off and used for animal feed

filter

nutrient solution containing penicillin

solvent in tank

penicillin dissolves in solvent

penicillin solution

chemicals in tank

penicillin reacts with chemicals to become crystals

crystals

penicillin crystals washed and dried to give final product

**Figure 8.18** *Purifying penicillin*

the process to a halt when the required level of antibiotic has been produced. This technology makes the system very efficient because it saves energy and prevents raw materials being wasted.

### Purifying the product

Figure 8.18 shows a simplified version of the method used to purify penicillin.

## Activity

## Constructing and using a simple fermenter

### You need

- dilute disinfectant (e.g. virkon)
- lemonade bottle (2 litre plastic)
- three size 9 rubber stoppers (one hole)
- thermometer
- lengths of plastic tubing
- diffuser stone
- two clips
- two litres of boiled and cooled sucrose solution (1%)
- packet of dried yeast (e.g. fast action Hovis or Allinson)
- foam or cotton wool bung
- pasco pH sensor attached to computer (if available)
- aquarium pump
- plasticine
- small beaker
- Clinistix strips

| factor being investigated | at start | after 2–3 days |
|---|---|---|
| glucose (present or absent) | | |
| temperature | | |
| pH | | |

**Table 8.3**  *Fermenter results*

### What to do

1  Make a copy of table 8.3.
2  Clean all parts of the equipment with disinfectant.
3  Set up the fermenter as shown in figure 8.19.
4  Add the sucrose solution and yeast.
5  Fit the foam bung (and pH sensor if available).
6  Turn on the aquarium pump and use the clip to adjust the air to a steady flow.
7  Plug any tiny leaks with plasticine if necessary.
8  Take a small sample of the fermenter's contents at the start via the sampling tube and test it for glucose using a Clinistix strip. Enter the result in your table.

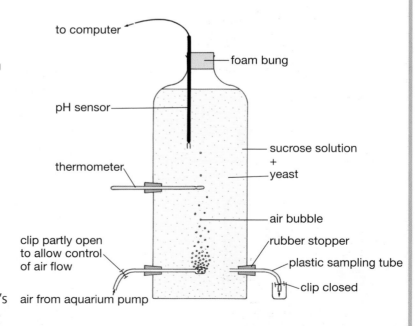

**Figure 8.19**  *A simple fermenter*

9  Read the starting temperature and pH and enter them in your table.
10 After about 2–3 days (maximum 4) test a sample of the fermenter's contents for glucose using a Clinistix strip. Enter the result in your table.
11 Take the temperature and pH and enter the results in your table.
12 Draw conclusions from your results.

## Environmental impact

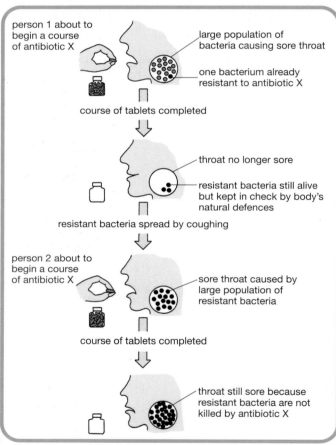

**Figure 8.20**  *Spread of resistant bacteria*

person 1 about to begin a course of antibiotic X

large population of bacteria causing sore throat

one bacterium already resistant to antibiotic X

course of tablets completed

throat no longer sore

resistant bacteria still alive but kept in check by body's natural defences

resistant bacteria spread by coughing

person 2 about to begin a course of antibiotic X

sore throat caused by large population of resistant bacteria

course of tablets completed

throat still sore because resistant bacteria are not killed by antibiotic X

### Development of resistance to antibiotics

In the early 1940s when penicillin became widely available for use, people thought that antibiotics would soon put an end to bacterial diseases. However as early as 1948, 75% of the *Staphylococcus* bacteria found amongst hospital patients and staff were **resistant** to penicillin.

Every time a new antibiotic is discovered, it is the same story. The drug is very effective at the start but soon resistant strains of bacteria appear, spread and increase in number. The antibiotic does not work against the resistant strain. Figure 8.20 shows in a simple way how resistant bacteria spread.

### Super Staph

*Staphylococcus aureus* (see figure 8.21) is a species of bacteria commonly found on the surface of human skin. It can cause boils and abscesses. In the past these were successfully treated using antibiotics. However there now exists a strain of *Staphylococcus aureus* known as '**Super *Staph***' which is resistant to *all but one* antibiotic.

### Over-use of antibiotics

Over the last 50 years, the production of antibiotics has rocketed (see figure 8.22).

### Medical applications

One of the reasons why strains of bacteria resistant to antibiotics have increased in number is because many doctors have over-

**Figure 8.21**  *Staphylococcus aureus*

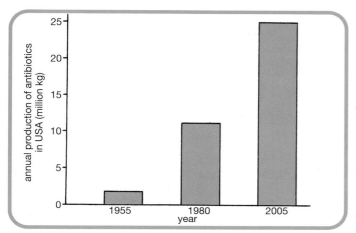

**Figure 8.22** *Increasing production of antibiotics*

**prescribed** antibiotics in recent years. This means that they have sometimes given patients a prescription for an antibiotic when the patient was not suffering from a serious infection and would have made a full recovery *without* the antibiotic. When antibiotics are over-used this leads to the death of more and more of the sensitive bacteria. Soon the resistant strains face hardly any competition and multiply freely.

## Down on the farm

Farm animals are often raised in crowded conditions (see figure 8.23) which allow easy

**Figure 8.23** *These chickens are developing in very crowded conditions*

spread of disease. Some farmers add **antibiotics** to the animal feed to prevent disease and allow the animals to grow rapidly. However this means that any bacteria that are **resistant** to the antibiotics will thrive and multiply. These could then be passed on to humans in **contaminated products** and lead to food poisoning. A person suffering food poisoning caused by bacteria resistant to antibiotics will *not* respond to treatment using these antibiotics. Experts believe that resistant strains of *Salmonella* and *Enterococcus* are now being transmitted from animals to humans (see figure 8.24).

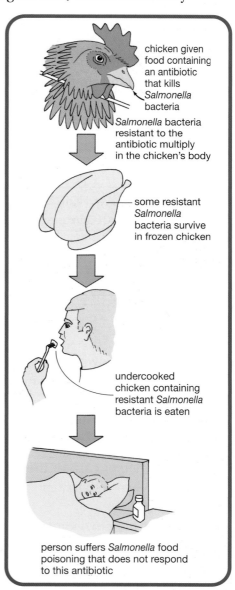

chicken given food containing an antibiotic that kills *Salmonella* bacteria

*Salmonella* bacteria resistant to the antibiotic multiply in the chicken's body

some resistant *Salmonella* bacteria survive in frozen chicken

undercooked chicken containing resistant *Salmonella* bacteria is eaten

person suffers *Salmonella* food poisoning that does not respond to this antibiotic

**Figure 8.24** *Transmission of resistant bacteria*

197

## Reducing the problem

There are several ways of tackling the problem of increasing resistance of bacteria to antibiotics. Some of these are as follows.

- Never take antibiotics when suffering a viral disease.
- Only take antibiotics on rare occasions when suffering a serious bacterial infection.
- Always finish the course of antibiotic treatment to try to kill all the bacteria and reduce the chance of some surviving and later becoming resistant.
- Support moves to keep new antibiotics for use only against infections that are resistant to other antibiotics.
- Support the EU ban on the use of antibiotics in animal feed.

## Testing your knowledge

**1** Copy and complete figure 8.25 using arrows to connect each term to its correct meaning. (3)

| | |
|---|---|
| industrial fermenter | system which allows conditions in a fermenter to be monitored and adjusted |
| genetically engineered microbe | enormous container in which fermentation is brought about by a microbe |
| computer-controlled technology | organism that has had its genetic material altered to make it useful |

**Figure 8.25**

**2** Rewrite the following sentences including only the correct answer in each choice in italics.
   a) Two types of micro-organism that are often genetically engineered are *bacteria/insulin* and *vaccines/yeast*.
   b) Two pharmaceutical products made by such genetically engineered micro-organisms are *bacteria/insulin* and *vaccines/yeast*. (4)

**3** Decide whether each of the following statements is true or false and use T or F to indicate your choice. When a statement is false, give the word that should have been used in place of the word in **bold** print. (6)
   a) One of the growing conditions monitored in a fermenter is **temperature**.
   b) Over-prescription of antibiotics can lead to bacteria becoming **sensitive** to antibiotics.
   c) When antibiotics are over-used, **resistant** bacteria multiply in the absence of competition.
   d) Some farmers add **antifungals** to the animal feed to prevent bacterial diseases.
   e) 'Super *Staph*' is a **fungus** that is resistant to all but one antibiotic.
   f) Antibiotics should not be taken to treat a **viral** infection.

## Applying your knowledge

**1** Figure 8.26 shows an experiment set up to investigate the growth of two microbes (X and Y) on nutrient agar. Each microbe was added to the nutrient agar using a wire loop.

a) Describe a method that could be used to sterilise the wire loop. (1)

b) Why was the Petri dish lid held in the position shown in figure 8.26? (1)

c) Why should the experiment be set up near a lit Bunsen burner? (1)

d) Which microbe showed more growth after several days? (1)

e) (i) Which microbe made a chemical substance that slowed down the growth of the other microbe?

(ii) Explain your answer to (i).

(iii) What general name is given to such a chemical? (3)

f) Which of the two microbes could be described as being *sensitive* to the other? (1)

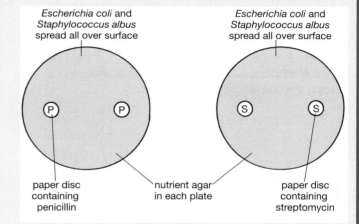

**Figure 8.27**

**2** The experiment shown in figure 8.27 was set up to investigate the effect of two antibiotics (penicillin and streptomycin) on the growth of two species of bacteria. State TWO ways in which the experiment must be altered to make it a valid test. (2)

**3** Read the passage on page 200 and answer the following questions.

a) What was the variable factor that was being tested in these experiments? (1)

b) Give ONE variable factor that was kept the same each time. (1)

c) (i) Which set of results was more reliable – those from the first or those from the second experiment?

(ii) Explain your answer. (2)

d) (i) Which part of each experiment was the control?

(ii) Why was the control experiment carried out? (2)

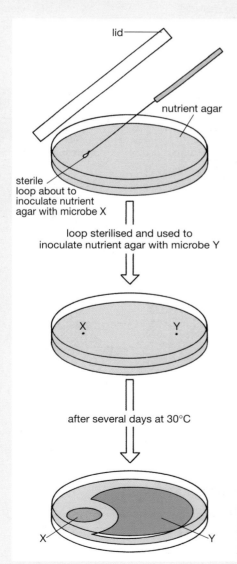

**Figure 8.26**

After Fleming made his famous discovery, another scientist continued the work on penicillin. He took eight mice and gave them a deadly dose of disease-causing bacteria (see figure 8.28). He treated four of them with penicillin and left the other four untreated. Within 24 hours the untreated mice had died while the treated mice were still alive.

Because the number of mice in the first trial was so small, he decided to repeat the experiment. This time he used 50 mice. He gave them all the deadly dose of bacteria and treated 25 mice with penicillin.

Again within 24 hours, all the untreated mice had died. Out of the 25 mice that were given the penicillin, 24 survived.

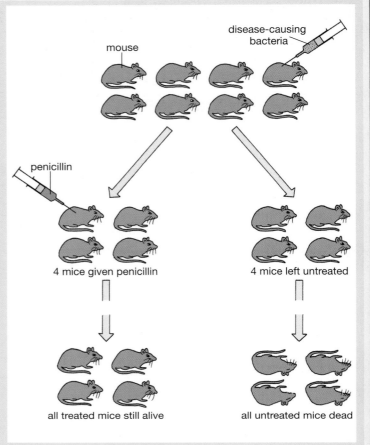

**Figure 8.28**

4 A doctor took a swab from a patient with a sore throat. A hospital laboratory technician spread the bacteria over the surface of sterile nutrient agar in a Petri dish. He took a multidisc and placed it on top of the bacteria. Figure 8.29 shows the result after 48 hours in a warm incubator.

a) (i)  Which antibiotics were able to prevent growth of the bacteria?

(ii) To which of these were the bacteria most sensitive?

(iii) Explain your answer to (ii). (3)

b) To how many of the antibiotics were the bacteria resistant? (1)

c) The patient was known to be allergic to antibiotics A, B and F.

(i)  Which antibiotic should be used to treat her?

(ii) Give a reason for your choice. (2)

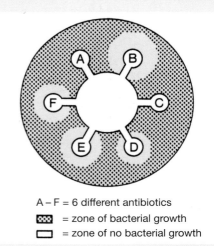

A – F = 6 different antibiotics
▨ = zone of bacterial growth
▢ = zone of no bacterial growth

**Figure 8.29**

d) The technician only set up one Petri dish. What should have been done to increase the reliability of the results? (1)

| bacterial species | antibiotic | | | |
|---|---|---|---|---|
| | W | X | Y | Z |
| 1 | + | + | − | − |
| 2 | − | + | + | − |
| 3 | − | + | + | − |
| 4 | + | − | − | + |

**key** + = antibiotic prevents growth of bacteria

− = antibiotic does not prevent growth of bacteria

**Table 8.4**

5 Table 8.4 shows the results of testing four antibiotics (W, X, Y and Z) on four species of bacteria (1, 2, 3 and 4).

a) Which antibiotic is able to stop three different species of bacteria from growing? (1)

b) Which TWO antibiotics are able to prevent growth of bacterial species 1? (1)

c) Which TWO antibiotics are unable to prevent growth of bacterial species 4? (1)

d) How many of the four species of bacteria are prevented from growing by antibiotic Y? (1)

e) Which of the four species of bacteria is sensitive to antibiotic Z? (1)

f) Which of the four species of bacteria are resistant to antibiotic W? (1)

6 The key in figure 8.30 refers to different medical treatments used against bacteria and fungi.

a) How many of the medical treatments in the key are antifungals? (1)

b) How many of the medical treatments in the key are produced by a type of microbe called *Streptomyces*? (1)

c) Which treatment would be used against each of the following?

(i) tuberculosis,

(ii) thrush,

(iii) bronchitis. (3)

d) Give TWO features that tetracycline and chloramphenicol have in common? (2)

e) Give TWO differences between penicillin and chloramphenicol. (2)

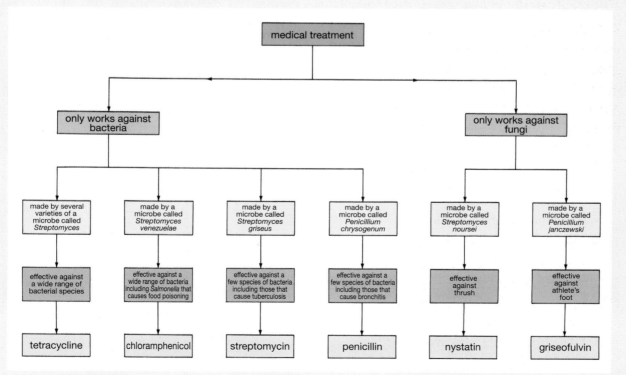

**Figure 8.30**

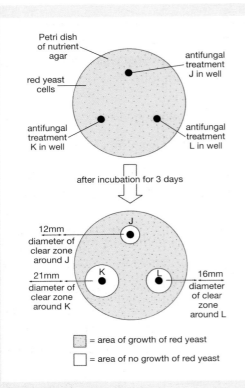

**Figure 8.31**

| procedure | reason |
|---|---|
| ① wash feet daily in warm soapy water | Ⓐ to prevent reinfection by spores left in sweaty socks |
| ② put fungicide powder between toes | Ⓑ to cut down the volume of sweat produced by the feet |
| ③ put clean socks on every day | Ⓒ to remove fragments of infected skin and spores from between toes |
| ④ wear leather rather than plastic footwear | Ⓓ to prevent bare feet picking up fungal spores from another person |
| ⑤ wear flip flops in communal shower areas | Ⓔ to create conditions that stop the growth of fungi |

**Figure 8.32**

8 The left-hand list in figure 8.32 gives five procedures that help to clear athlete's foot and stop it coming back.
   a) Match each of these with the reason for carrying it out given in the right hand list. (5)
   b) During the war in Iraq in 2003, many British and American soldiers on active duty in the desert for several weeks suffered a severe attack of athlete's foot. Suggest why. (2)

9 Figure 8.33 shows eight Petri dishes set up to investigate the effect of two antifungal chemicals (X and Y) under different conditions.
   a) What variable factor can be investigated by comparing the results of dishes 1 and 3? (1)
   b) What variable factor can be investigated by comparing the results of dishes 2 and 6? (1)
   c) Which TWO dishes should be compared to find out the effect of the type of antifungal on red yeast at 40°C? (1)
   d) Which TWO dishes should be compared to find out the effect of temperature on the action of antifungal X on brown yeast? (1)
   e) Which TWO dishes should be compared to find out which colour of yeast is more severely affected by antifungal Y at 40°C? (1)

7 Figure 8.31 shows an investigation set up to find out how effective three antifungal treatments (J, K and L) were at preventing red yeast from growing. Red yeast cells were spread evenly over the surface of the nutrient agar and then three wells were cut in the agar. An antifungal cream was placed in each well and the dish incubated at 30°C for 3 days.
   a) Draw a bar chart of the results. Put diameter of clear zone on the vertical (y) axis. (3)
   b) What was the variable factor under investigation in this experiment? (1)
   c) (i) Which antifungal treatment was best at preventing red yeast cells from growing?
      (ii) Explain your answer. (2)
   d) What control should have been carried out? (1)
   e) What could be done to increase the reliability of the results? (1)

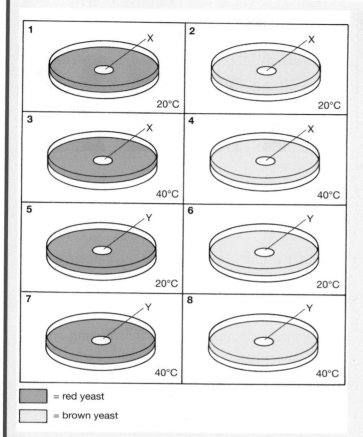

= red yeast

= brown yeast

**Figure 8.33**

| time (hours) | number of bacteria |
|---|---|
| 0 | 1 |
| 0.5 | 2 |
| 1 | W |
| 1.5 | 8 |
| 2 | 16 |
| 2.5 | X |
| 3 | 64 |
| 3.5 | Y |
| 4 | 256 |
| 4.5 | 512 |
| 5 | Z |
| 5.5 | 2048 |
| 6 | 4096 |

**Table 8.5**

**10** Table 8.5 shows the results of growing a population of bacteria from a single cell in a fermenter for 6 hours. Under ideal growing conditions, the number doubles every half hour.
   a) State the number of bacteria that should have been given at boxes W, X, Y and Z in the table. (4)
   b) There were 16 bacteria at hour 2. By how many times had this number increased at hour 6? (1)
   c) Predict the number of bacteria that would be present at 7 hours if the same trend continued. (1)
   d) (i)   Name TWO conditions in an industrial fermenter that would need to be monitored to make sure that they stayed at a level that suited the bacteria.
       (ii)  What equipment is used to monitor and control these conditions?
                                                   (3)

**11** Table 8.6 shows the results of growing a microbe in a fermenter to produce an antibiotic called streptomycin.
   a) Draw a line graph of the results with sampling time on the horizontal (x) axis. (4)
   b) Between which TWO sampling times did the microbe start to make streptomycin? (1)
   c) Between which TWO sampling times did the microbe make the antibiotic at the fastest rate? (1)
   d) By how many times was the concentration of streptomycin present at 80 hours greater than that at 40 hours? (1)
   e) Express the concentration of streptomycin at 120 hours compared with that at 40 hours as a whole number ratio. (1)

| sampling time (hours) | concentration of streptomycin (units) |
|---|---|
| 0 | 0 |
| 20 | 0 |
| 40 | 3 |
| 60 | 15 |
| 80 | 21 |
| 100 | 24 |
| 120 | 24 |

**Table 8.6**

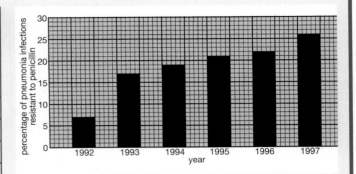

**Figure 8.34**

**12** The bar graph in figure 8.34 shows the results from a survey on the percentage of pneumonia infections in USA that were resistant to penicillin.

a) What percentage of pneumonia infections showed resistance to penicillin in
(i) 1993? (ii) 1997? (2)

b) What overall trend is shown by the graph? (1)

c) Between which TWO years did the greatest increase in percentage of pneumonia infections resistant to penicillin occur? (1)

d) By how many times did the percentage of pneumonia infections resistant to penicillin increase between 1992 and 1995? (1)

e) If a patient is found to be suffering an infection resistant to penicillin, what treatment would the doctor give next? (1)

f) State TWO things that people can do to try to help to slow down the increase in number of bacteria resistant to antibiotics. (2)

## What you should know

| | | |
|---|---|---|
| adjusted | Fleming | resistance |
| antibiotic | fungi | resistant |
| antifungal | genetic | rivals |
| athlete's foot | growth | sensitive |
| bacteria | less | slows |
| desired | micro-organism | temperature |
| destroys | monitored | thrush |
| engineered | over-prescription | vaccine |
| farmers | penicillin | viruses |
| fermenters | products | yeast |

**Table 8.7**    *Word bank for chapter 8*

1  An _____ is a chemical substance produced naturally by one type of micro-organism (e.g. a fungus) that _____ or prevents the further growth of another type of _____ (e.g. bacterium). Antibiotics do not work on _____.

2  Different antibiotics are effective against different bacteria. If the bacterium's growth is prevented by the antibiotic, the bacterium is said to be _____; if the antibiotic has no effect, the bacterium is said to be _____.

3  Antibiotics are produced naturally by soil _____ to kill their _____.

4  The first antibiotic was discovered by Alexander _____. He called it _____.

5  An _____ is a chemical substance that _____ down or stops the growth of fungal infections such as _____ and _____.

6  The technology that allows scientists to transfer genes from one living organism to another is called _____ engineering. This procedure makes it possible for scientists to create micro-organisms that have been genetically _____ to produce a _____ pharmaceutical product (e.g. _____ cells that can make hepatitis B _____).

7  The pharmaceutical industry grows useful micro-organisms on a vast scale in industrial _____ to produce huge quantities of _____ such as antibiotics.

8  In an industrial fermenter, conditions such as _____ are carefully _____ by computers and are automatically _____ when necessary to give the microbe the ideal conditions for

_____.

9  When a new antibiotic is discovered, it is normally very effective at the start but soon becomes _____ effective as resistant strains of _____ appear and increase in number.

10  _____ of antibiotics by doctors and over-use of antibiotics by _____ can both lead to an increase in number of bacteria that develop _____ to antibiotics.

# Growing Plants

Growing plants from seeds or by vegetative propagation and then maintaining them in a healthy environment enables people to produce plants on a massive scale for profit.

# 9 Growing plants from seeds

## Seed biology

Flowering plants make **seeds** to *reproduce* themselves. Each seed contains all the parts needed to start the growth of a new plant.

### Seed structure

A seed is made up of an **embryo** and a **food store** surrounded by a **seed coat**. Figure 9.1 shows the structure of a broad bean seed. Table 9.1 gives the functions of the different parts.

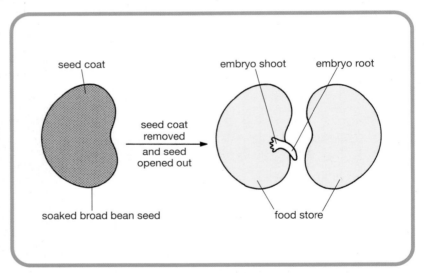

**Figure 9.1** *Structure of a broad bean seed*

| part of seed | function of part |
|---|---|
| embryo root | to grow into the plant's root |
| embryo shoot | to grow into the plant's shoot |
| food store | to provide energy for growth of the embryo root and shoot |
| seed coat | to protect the seed from damage by soil micro-organisms |

**Table 9.1** *Functions of seed parts*

## Activity

### Examining the structure of bean and pea seeds

**You need**

- small cork board ('coaster')
- three pins
- soaked broad bean seed
- soaked pea seed
- soaked mung bean seed
- dropping bottle of iodine solution
- hand lens

| name of seed type | starch present in food store? (yes/no) | number of tiny leaves in embryo shoot |
| --- | --- | --- |
| | | |
| | | |
| | | |

**Table 9.2** *Seed dissection results*

**What to do**

1. Make a copy of results table 9.2.
2. Examine the broad bean seed by peeling off its tough seed coat and opening out its contents.
3. Use two pins to anchor to the cork board one half of the bean with its inside surface upwards.
4. Using the third pin, scratch the inside surface of the bean a few times in a criss-cross pattern.
5. Add three drops of iodine solution to the scratched surface to test for the presence of starch. (A blue-black colour indicates a positive result.) Enter the result in your table.
6. Examine the embryo root and shoot with the aid of the hand lens.
7. Count how many tiny leaves are present in the embryo shoot using a pin and enter the result in your table.
8. Repeat steps 2–7 using a pea seed.
9. Repeat steps 2–7 using a mung bean seed.
10. Compare your results with figure 9.2.
11. Draw conclusions from your results.

**Figure 9.2** *Dissection of seeds*

# Seeds from other plants

Seeds come in many shapes and sizes. Sometimes they are found inside a fruit such as a **berry**, a **nut** or a **cereal grain**. Figures 9.3 and 9.4 show examples of seeds and fruits.

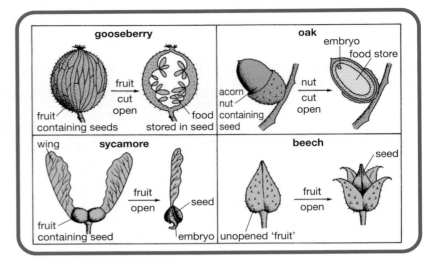

**Figure 9.3**   *Seeds and fruits (from bushes and trees)*

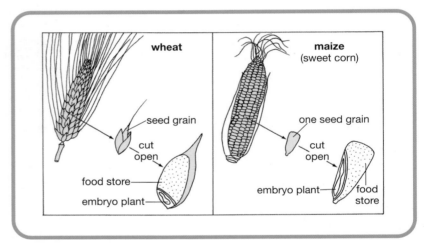

**Figure 9.4**   *Seeds and fruits (from cereal plants)*

**Activity**

## Examining seeds from a variety of plants (see also figure 9.5)

You need
- ceramic tile
- pin
- forceps
- hand lens
- soaked seeds (e.g. sweet pea, sunflower, mustard)
- berries, whole and halved (e.g. gooseberry, elderberry, rowan)
- tree seeds (e.g. sycamore, beech, oak)
- additional examples, whole and halved (e.g. wheat, maize, peanut)
- dropping bottle of iodine solution

## What to do

1 Using forceps and a pin, carefully remove the seed coat from a soaked seed.
2 Using a hand lens, examine the contents of the seed and locate the embryo root and shoot.
3 Add three drops of iodine solution to the seed's food store to test for starch.
4 Carefully remove the outer coat from the seed of a tree and repeat steps 1–3 above.
5 Using the same procedure, examine the structure of whole and halved berries and other available examples.

**Figure 9.5**   *Examining seeds and fruits*

Activity

## Measuring the water content of stored and fresh pea seeds

### You need

- two sheets of aluminium baking foil (approx 300 mm × 200 mm)
- 100 stored ('dry') pea seeds
- 100 pod fresh (or defrosted frozen) pea seeds
- access to an electronic balance
- access to an oven set at 80°C
- permanent felt tip marker

### What to do

1 Make a copy of table 9.3.
2 Collect the two sheets of baking foil and make them into 'oven dishes' as shown in figure 9.6 on page 212.
3 Using the marker, add your initials to each dish. Mark one of them 'stored' and the other 'fresh' (see figure 9.7).

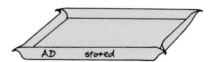

**Figure 9.7**   *Oven dish*

| | | type of pea seed | |
|---|---|---|---|
| | | stored | fresh |
| a) | mass of empty oven dish (g) | | |
| b) | mass of dish + 100 seeds before drying (g) | | |
| c) | mass of 100 seeds before drying (g) [b) − a)] | | |
| d) | mass of dish + 100 seeds after drying (g) | | |
| e) | mass of water lost by 100 seeds (g) [b) − d)] | | |
| f) | Percentage water content of seeds [e)/c) × 100] | | |

**Table 9.3**   *Results for water content of seeds*

211

4  Weigh the 'stored' dish empty and enter the result in your table.

5  Weigh the 'stored' dish containing 100 stored pea seeds and enter the result in your table.

6  Repeat steps 4 and 5 for fresh pea seeds.

7  Place the two dishes of pea seeds in the oven at 80°C for two days.

8  Allow the dishes to cool down before weighing them.

9  Repeat steps 7 and 8 until there is no further change in mass and all of the water in the seeds has been removed.

10  Use these final results to complete your table.

11  Draw a conclusion from the experiment about the percentage water content of stored and fresh pea seeds.

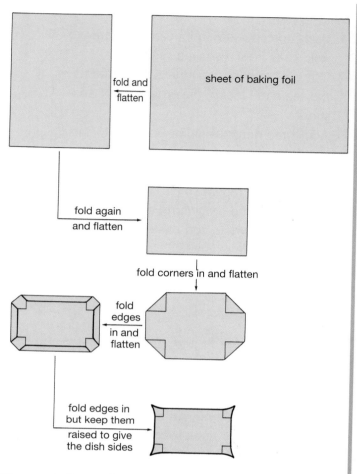

**Figure 9.6**  *Making an oven dish*

## Learning about three food tests

You need

- Clinistix strips
- Albustix strips
- dropping bottle of iodine solution
- dropping bottle of starch (1% suspension)
- dropping bottle of glucose (1% solution)
- dropping bottle of protein (1% suspension of egg albumen)
- nine test tubes
- test tube stand to take nine test tubes
- permanent felt tip marker

What to do

1 Make a copy of results table 9.4.
2 Use the marker to label the test tubes 1–9.
3 Add 30 drops of starch to tubes 1, 2 and 3.
4 Add 30 drops of glucose to tubes 4, 5 and 6.
5 Add 30 drops of protein to tubes 7, 8 and 9.
6 Add a Clinistix strip to tubes 1, 4 and 7 to find out which type of food it gives a positive result with. (This is indicated by a change in colour from pink to purple.) Enter the result in your table.
7 Add an Albustix strip to tubes 2, 5 and 8 to find out which type of food it gives a positive result with. (This is indicated by a change from yellow to green.) Enter the result in your table.
8 Add three drops of iodine solution to tubes 3, 6 and 9 to find out which type of food it gives a positive result with. (This is indicated by a change from brown to blue-black.) Enter the result in your table.

| food test | positive result | type of food present |
|---|---|---|
| Clinistix | | |
| Albustix | | |
| iodine solution | | |

**Table 9.4** *Three food tests*

## Activity

# Testing seeds for the three types of food

### You need
- three test tubes labelled 1, 2 and 3
- pestle (grinder) and mortar (bowl)
- variety of seeds (e.g. pea, broad bean, sunflower, mung bean, mustard, sweet corn, wheat)
- test tube stand
- Clinistix strips
- Albustix strips
- dropping bottle of iodine solution
- measuring cylinder (10 cm³ plastic)

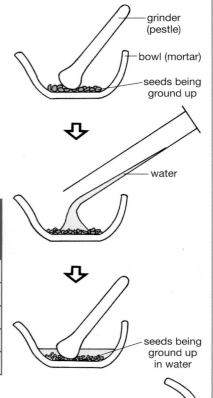

| seed type | glucose test (Clinistix) | protein test (Albustix) | starch test (Iodine solution) |
|---|---|---|---|
| cabbage | – | + | – |
| | | | |
| | | | |
| | | | |
| | | | |

**Table 9.5**   *Food test results for seeds*

### What to do (also see figure 9.8)

1 Prepare a table of results as shown in table 9.5 (+ = positive result showing that the food is present; – = negative result showing that the food is absent). Cabbage has been given as an example.

2 Grind a few of the first type of seed (e.g. pea) in the mortar using the pestle.

3 Using the measuring cylinder, add 5 cm³ of water and grind again.

4 Pour an equal volume of the liquid into test tubes 1, 2 and 3.

5 Add a Clinistix strip to tube 1 and enter + in your table if glucose is present and – if glucose is absent.

6 Add an Albustix strip to test tube 2 and enter + in your table if protein is present and – if protein is absent.

7 Add three drops of iodine solution to test tube 3 and enter + in your table if starch is present and – if starch is absent.

8 Clean out the mortar as instructed by your teacher and rinse out the three test tubes thoroughly.

9 Repeat steps 2–8 with each type of seed available.

10 Draw a conclusion from your results.

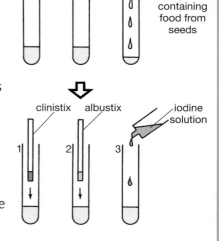

**Figure 9.8**   *Testing seeds for three types of food*

# Germination

Germination is the growth of a plant embryo into a plant with green leaves. During germination, the growing embryo uses the reserves of food stored in the seed. Figures 9.9, 9.10, 9.11 and 9.12 show germination in four different plants.

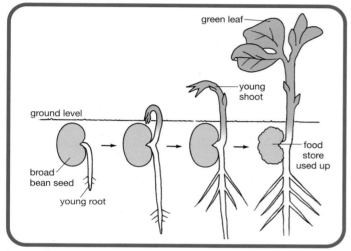

**Figure 9.9**  *Germination in a broad bean*

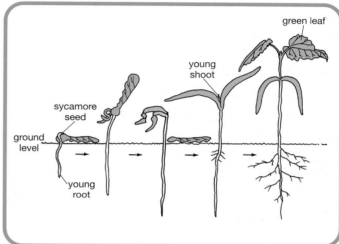

**Figure 9.10**  *Germination in sycamore*

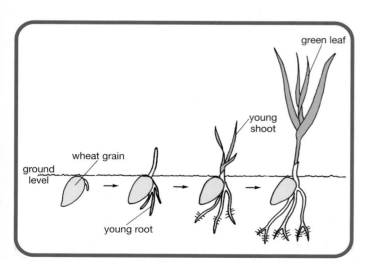

**Figure 9.11**  *Germination in wheat*

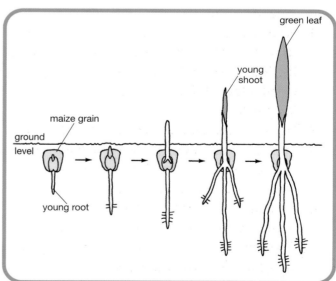

**Figure 9.12**  *Germination in maize*

## Rules when carrying out a biological investigation

In biology, the results of an investigation are valid if:

- at each stage only **one variable factor** is studied at a time. If several were involved then it would be impossible to know which factor was responsible for the results.
- a **large number** of the organism being investigated is used. If only a few were used, they could be unusual in some way and not behave normally.
- the experiment can be successfully **repeated many times**. If not then perhaps the result just came about by chance and is not reliable.

### Activity

## Investigating the conditions necessary for germination of seeds

### You need

- four test tubes
- cotton wool
- cress seeds
- 100 cm³ of boiled and cooled water
- dropping bottle of oil
- access to freezer

### What to do

**1** Set up the experiment as shown in figure 9.13. This satisfies the three rules given above for the following reasons:
  - tubes B, C and D each differ from tube A by only **one variable factor**;
  - the same **large number** of cress seeds is used in each tube to allow for some being unusual or dead;
  - the whole experiment is repeated by several groups of students.

**2** Make a copy of table 9.6 and complete rows 1–3.

**3** Examine the test tubes after 4 days.

**4** Complete row 4 in the table to show your results.

**5** Draw a conclusion about the conditions necessary for the germination of cress seeds.

A — tube at room temperature

B — tube at room temperature

C — tube at room temperature

D — tube in freezer

50 cress seeds in each tube

oil

cotton wool soaked with tap water

cotton wool soaked with boiled and cooled tap water lacking oxygen

dry cotton wool

cotton wool soaked with tap water

**Figure 9.13**  *Investigating the conditions needed for germination*

| | tube | | | |
|---|---|---|---|---|
| | **A** | **B** | **C** | **D** |
| **1** oxygen present or absent? | | | | |
| **2** water present or absent? | | | | |
| **3** room temperature or freezing temperature? | | | | |
| **4** germination or no germination? | | | | |

**Table 9.6** *Germination conditions and results*

# Plant life cycle

The series of changes that a plant goes through as it develops from a certain stage (e.g. germinating seed) until it reaches the same stage in the next generation is called its **life cycle**. The stages in a plant's life cycle can be observed within a few weeks by growing *Brassica* 'fast plants'.

## Activity

### Assembling a diagram of the life cycle of a 'fast plant'

*You need*
- copy of figure 9.14 (parts 1 and 2)
- scissors
- glue stick

*What to do*
1 Cut out the five parts of the plant life cycle shown in part 1 of figure 9.14.
2 Glue them on to the 'wheel' in part 2 in the correct sequence. (The first stage of the plant life cycle is already in place.)

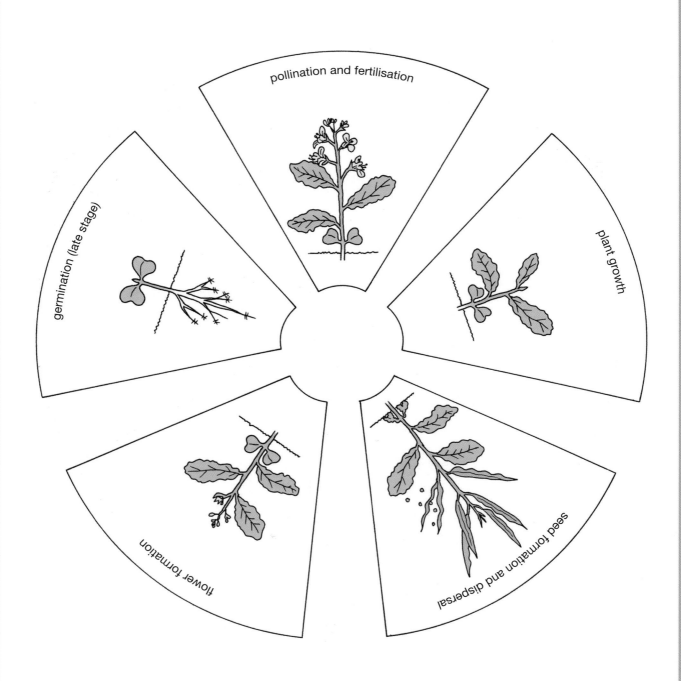

pollination and fertilisation

plant growth

germination (late stage)

flower formation

seed formation and dispersal

**Figure 9.14 Part 1** *Plant life cycle*

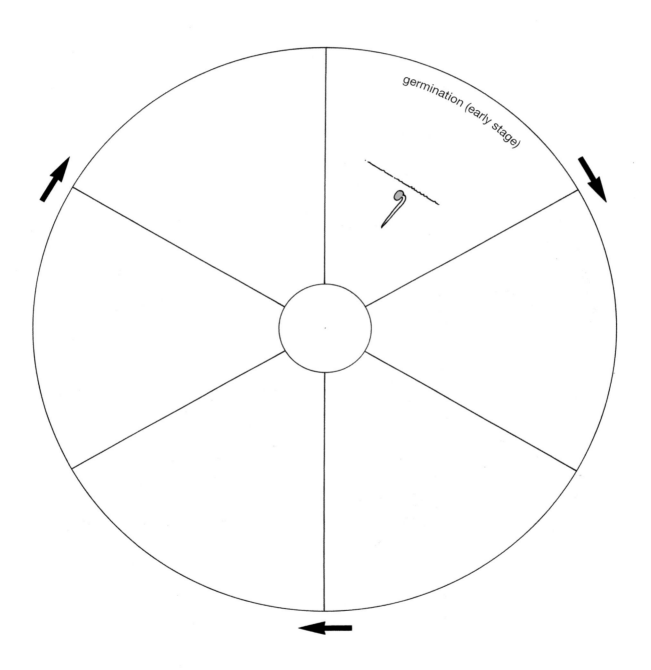

germination (early stage)

**Figure 9.14 Part 2**  *Plant life cycle*

## Dormancy

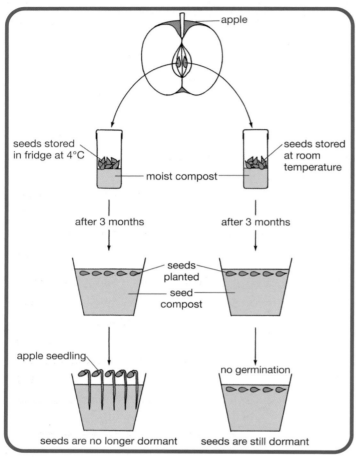

**Figure 9.15**  *Breaking dormancy in apple seeds*

The seeds of many plants are released into the surrounding environment during summer or autumn. Many of these seeds may land in places with all of the conditions necessary for germination:

- water
- oxygen
- suitable temperature.

However the seeds of a great number of plant types do *not* germinate until the following spring. These seeds, which are alive but inactive, are said to be **dormant**.

In some types of seed (e.g. conifer trees, apple and mountain ash) the dormancy can be **broken** by exposing the seeds to a low temperature (e.g. 4°C) for 2–3 months. By this means the seeds have been exposed artificially to *winter-like* conditions (see figure 9.15).

Under natural conditions, a period of dormancy in seeds during the winter is of advantage because it **delays germination** until favourable growing conditions arrive in spring when the soil temperature increases. If the seeds germinated in autumn, the young seedlings produced would probably be killed by winter frost.

## Testing your knowledge

1  a) Why do plants make seeds?  (1)
   b) (i)   Once in the ground, what would happen to a seed's embryo plant and food store if they were not surrounded by a seed coat?
      (ii)  What would happen to an embryo plant in a seed if it began to grow and did not have a food store?  (2)

2  Copy and complete table 9.7 which refers to three food tests.  (6)

3  a) What does the word *germination* mean?  (2)
   b) (i)   Draw a labelled diagram of an experiment (using only two test tubes) to show that water is needed by cress seeds to germinate.

| type of food being tested for | name of chemical used for test | colour of chemical before use | colour resulting if food is present |
|---|---|---|---|
| starch | iodine solution | | |
| glucose | | pink | |
| protein | | | green |

**Table 9.7**

    (ii)  Normally when a biological investigation is carried out, there are three rules to consider. Which TWO rules have been followed in your design?

    (iii) How could the experiment be extended to include the third rule?   (5)

  c) In addition to water, what TWO other conditions do seeds normally need to germinate?   (2)

**4** a) What is meant when a seed is described as being in a *dormant* state?   (1)

  b) What is the natural advantage to a seed of a period of dormancy during winter?   (1)

  c) By what means can dormancy in apple and mountain ash seeds be broken artificially?   (1)

# Photosynthesis

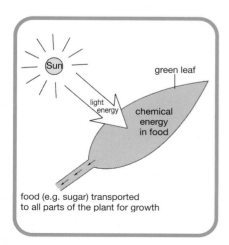

**Figure 9.16**  *Photosynthesis*

**Photosynthesis** is the process by which green plants make food. They do this by changing **light** energy into **chemical energy** contained in food (e.g. sugar) as shown in figure 9.16.

During germination, before photosynthesis begins, the embryo plant uses the food stored in the seed for growth. By the time these reserves run out, the young plant has developed *green leaves*. It is now able to *photosynthesise* and make its own food for growth. It has therefore become *independent*.

## Measuring changes in mass of germinating seeds and photosynthesising seedlings

Five groups of mung bean seeds (A–E) each weighing 20 g are used to set up the experiment shown in figure 9.17. The seeds in group A are dried in an oven at 80°C until they have a constant mass (their **dry mass**).

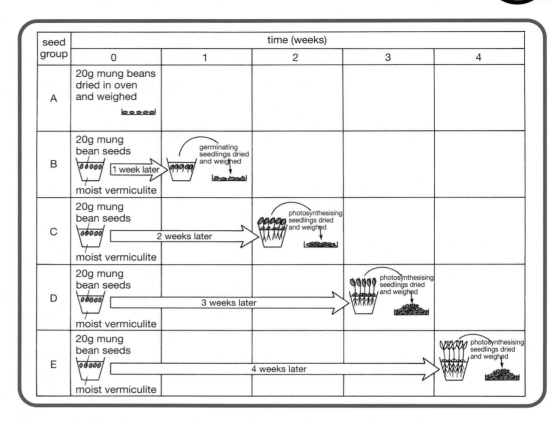

**Figure 9.17** *Measuring changes in mass of germinating seeds and photosynthesising seedlings*

Groups B, C, D and E are planted in pots of moist vermiculite (artificial soil) and kept in a seed propagator at 24°C. After one week the germinating seeds in group B are removed from their pot and carefully rinsed to remove all traces of vermiculite. They are then dried to constant mass in the oven to find their dry mass. This procedure is repeated week after week with group C, then group D and finally group E.

When this experiment is carried out, the results show that germinating seeds (those in group B) have *lost* mass when compared to the seeds in group A (the control group). This is because food stored in the seeds has been used by the embryo bean plants in B as a source of energy required for growth.

The results also show that the photosynthesising seedlings (those in groups C, D and E) have *gained* mass compared to the seeds in group B. This is because the bean seedlings now have green leaves which can photosynthesise and allow the growing plant to make its own food.

# Growing plants from seeds

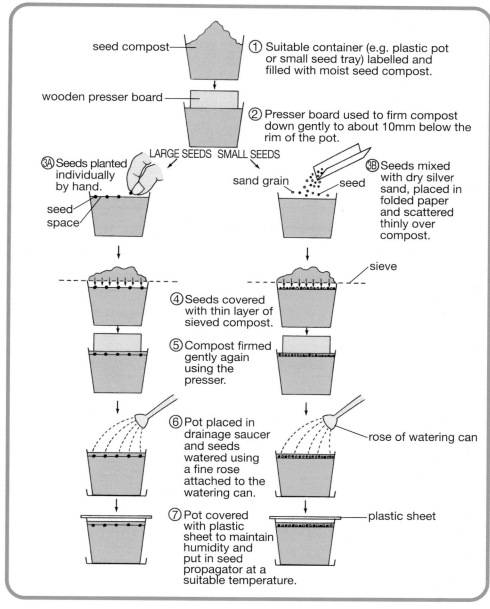

seed compost

① Suitable container (e.g. plastic pot or small seed tray) labelled and filled with moist seed compost.

wooden presser board

② Presser board used to firm compost down gently to about 10mm below the rim of the pot.

LARGE SEEDS   SMALL SEEDS

③A Seeds planted individually by hand.

sand grain   seed

③B Seeds mixed with dry silver sand, placed in folded paper and scattered thinly over compost.

seed space

sieve

④ Seeds covered with thin layer of sieved compost.

⑤ Compost firmed gently again using the presser.

⑥ Pot placed in drainage saucer and seeds watered using a fine rose attached to the watering can.

rose of watering can

⑦ Pot covered with plastic sheet to maintain humidity and put in seed propagator at a suitable temperature.

plastic sheet

**Figure 9.18** *Methods of spacing out seeds during sowing*

## Methods of seed sowing

To **sow** seeds means to put them in a place where they are likely to grow (e.g. moist soil). Seeds need to be *spaced out* during sowing to prevent overcrowding following germination.

Figure 9.18 shows two methods of spacing out seeds during sowing. Large seeds should be sown individually **by hand**. Small seeds should be mixed with a little **silver sand** before sowing. This diagram only gives a general broad outline. Different seeds have different requirements. For example at stage 4, although most types of seed should be just covered with a layer of sieved compost, *very* small seeds should *not* be covered. So it is important to read the instructions on the seed packet carefully.

## Activity

# Sowing seeds (assessment)

You need

- supply of seeds (either large seeds such as mung bean or small seeds such as pansy)
- label
- plastic plant pot (e.g. 100 mm top diameter)

- drainage saucer
- wooden presser board (e.g. 45 mm × 45 mm × 60 mm)
- seed compost
- dry silver sand
- sieve
- small watering can with fine rose
- plastic sheet (e.g. 120 mm × 120 mm)
- access to seed propagator

### What to do

**1** Decide whether the seeds that you have been given to sow are the large type or the small type.

**2** Consider figure 9.18 and decide which route is correct for your size of seeds.

**3** Carefully carry out the procedure.

**4** Before placing the pot of sown seeds in the propagator, show it to your teacher for assessment.

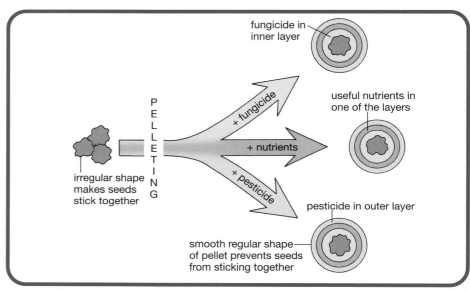

**Figure 9.19**  *Useful chemicals in pellets*

### Pelleted seeds

A **pelleted** seed is one that has been coated with several layers of material such as **clay**. These layers make up the pellet. The layers are applied in a tumbling machine. Seeds are pelleted for the following reasons.

### Regular seed shape

Pelleting makes irregularly shaped seeds **round** and **smooth**. They therefore do not tend to stick together and can be easily planted. Planting is often carried out using a machine called a drill that spaces the seeds out evenly.

### Application of chemicals

Different types of chemical (see figure 9.19) can be included amongst the layers in a pellet as follows:

- **fungicide** can be included in the inside layer to kill fungal spores normally carried on the surface of the seed. If left untreated, the fungal spores might cause disease to the young plant.
- **pesticide** can be put into the outer layer of the pellet to kill pests in the soil but leave the seed unaffected.
- **nutrients** can be included in one of the layers of the pellet. These encourage the growth of the young plant following germination of the seed.

Many different crop plants have their seeds pelleted, particularly *small* seeds such as sugar beet, carrot, lettuce, onion and tomato. Pelleting is rarely necessary for large seeds such as bean and pea because they are more easily handled.

Pelleted seeds need more water than unpelleted seeds and take longer to germinate. However pelleted seeds produce a better crop because the plants are better spaced out and healthier as a result of the useful chemicals present in the layers of the pellet.

## Activity

## Comparing germination in pelleted and unpelleted seeds

### Your need
- 100 pelleted seeds (e.g. carrot)
- 100 unpelleted seeds (same variety as before)
- two labels
- two small plastic plant pots (e.g. 80 mm top diameter)
- two drainage saucers
- wooden presser board (e.g. 45 mm × 45 mm × 60 mm)
- seed compost
- dry silver sand
- sieve
- small watering can with fine rose
- two plastic sheets (e.g. 100 mm × 100 mm)
- access to seed propagator

### What to do
1 Label the pots 'pelleted' and 'unpelleted' and add your initials.
2 Study figure 9.18.
3 Follow the procedure for *small* seeds.
4 Inspect your pots weekly to compare the percentage germination of 'pelleted' and 'unpelleted' seeds.
5 If possible, pool your results with those of other students.
6 Answer the following questions:
   a) What was the one variable factor studied in this experiment?
   b) Draw a conclusion about the effect of this factor on the germination of the seed type that you used.
   c) Why were as many as 100 seeds used each time?

**Figure 9.20** *Pre-soaked seeds*

**Figure 9.21** *A 'nicked' seed*

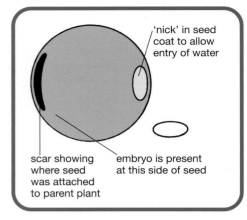

**Figure 9.22** *Where to 'nick' a seed*

## Pre-germination

**Pre-germinating** (chitting) a seed means making it start to germinate before it has been planted. This can often be done by soaking the seeds in water for a day or two. This is called **pre-soaking** (see figure 9.20).

Some seeds, however, have *very hard* seed coats which prevent water from entering easily and triggering the process of germination. Under natural conditions such seeds often do not germinate until part of their seed coat has been decomposed by decay micro-organisms in the soil, allowing water to get in. Seeds with very hard coats can be made to germinate more quickly by treating their seed coats in one of the following ways:

- If the seed is large, it may already have been *cracked* open by a threshing machine during harvesting. If not, it can be **slit** open or **'nicked'** using a sharp knife (see figure 9.21). This is best done at the side of the seed opposite to where the embryo is found and where the seed was once attached to the parent plant. This procedure avoids the risk of damage to the embryo (see figure 9.22).
- If the seed is large, a small portion of its hard seed coat can be **filed** away using a nail file.
- If the seeds are small (e.g. geranium), they can be **rubbed with abrasive paper** (e.g. emery paper) to break down their hard coats. This is best done by rolling them between two sheets of emery paper or shaking them in a closed container lined with abrasive paper.

Each of these methods breaks open the hard seed coat allowing water to enter and speed up germination.

## Activity

### Investigating the effect of chitting on germination of pea seeds

You need
- five chitted pea seeds (made to start to germinate by pre-soaking for four days)
- five unchitted pea seeds
- two labels
- two plastic plant pots (e.g. 100 mm top diameter)
- two drainage saucers
- wooden presser board (e.g. 45 mm × 45 mm × 60 mm)
- seed compost
- sieve
- small watering can with fine rose
- two plastic sheets (e.g. 120 mm × 120 mm)
- access to seed propagator

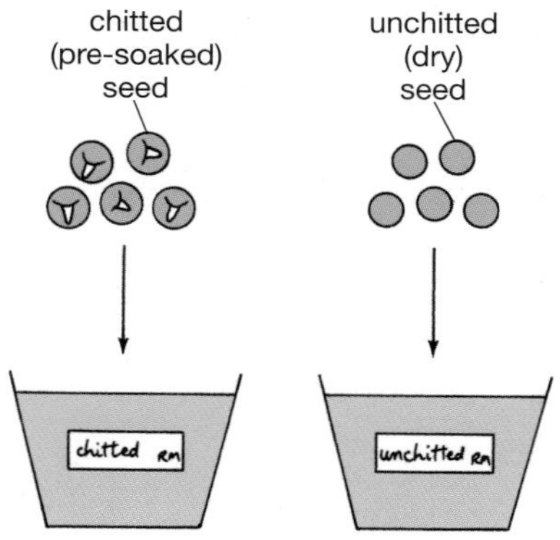

**Figure 9.23** *Investigating the effect of chitting*

What to do (also see figure 9.23)

1 Label the pots 'chitted' and 'unchitted' and add your initials.
2 Follow the procedure for *large* seeds shown in figure 9.18.
3 Inspect your pots weekly to compare the number of seeds that have germinated and developed into seedlings in each pot.
4 If possible, pool your results with those of other students.
5 a)   Draw a conclusion about the effect of chitting on the germination of pea seeds.
   b) State THREE factors that were kept the same so that a valid comparison could be made at the end of the experiment.

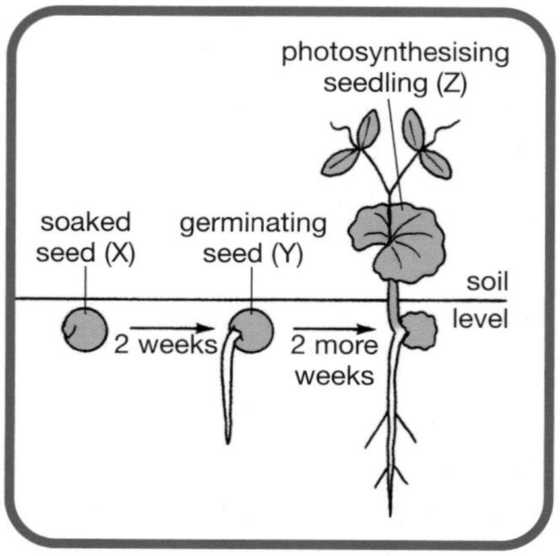

**Figure 9.24**

## Testing your knowledge

1 a) Describe what happens in a green plant during the process of photosynthesis. (2)
   b) Explain why photosynthesis is important to a plant. (1)

2 Figure 9.24 shows the development of a pea seedling from a seed.
   a) (i)  Compared with stage X, which stage (Y or Z) would have the *lighter* dry mass?
      (ii) Explain your answer (2)
   b) (i)  Compared with stage X, which stage (Y or Z) would have the *heavier* dry mass?
      (ii) Explain your answer. (2)

3 a) Why do seeds need to be spaced out during sowing? (1)
   b) (i)  Give ONE way of making sure that large seeds are well spaced out during sowing.

(ii) What different technique can be used to space out small seeds during sowing? (2)

4 Decide whether each of the following statements is true or false and use T or F to indicate your choice. Where a statement is false, give the word that should have been used in place of the word in **bold** print. (6)

a) Pelleted seeds have been coated with clay to give them an **irregular** shape.

b) Pelleted seeds normally take **less** time to germinate than unpelleted seeds.

c) Pelleted seeds are surrounded by layers that may contain chemicals such as **fungicide**.

d) Large seeds with very hard seed coats can be prepared for pre-germination by slitting them open using a **knife**.

e) Tiny seeds with very hard seed coats can be prepared for pre-germination by rubbing them between two sheets of **filter** paper.

f) To start germination before planting, seeds are **pre-soaked** in water for a day or two.

## Applying your knowledge

1 Figure 9.25 shows some of the stages in the development of a horse chestnut tree (diagram not drawn to scale).

a) Arrange stages A–H into a sequence that shows the life cycle of this plant starting with D. (1)

b) Redraw stage F and label the parts *seed coat*, *embryo root* and *embryo shoot*. (4)

c) The structure shown at stage D contains a store of food. Explain why. (1)

2 Table 9.8 shows the results from an experiment set up to compare the percentage water content of stored and fresh pea seeds.

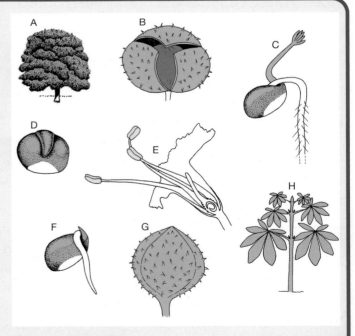

**Figure 9.25**

| | type of seed | |
|---|---|---|
| | stored | fresh |
| mass of 50 seeds (g) | 15 | 20 |
| mass of 50 dried seeds (g) | 12 | 8 |
| mass of water lost by 50 seeds | 3 | Y |
| percentage water content of 50 seeds | X | 60 |

**Table 9.8**

a) Why were as many as 50 of each type of seed used? (1)
b) During the experiment, the seeds were placed in an oven at 90°C for 24 hours and then weighed. This procedure was repeated day after day until there was no further change in their mass. What was the reason for carrying out this procedure? (2)
c) (i)  What was the percentage water content of the stored pea seeds (box X in table 9.8)?
   (ii) What was the mass of water lost by 50 fresh pea seeds (box Y in table 9.8)? (2)

**3** In an investigation, a student made use of the two food tests given in table 9.9. The student wanted to find out which type of food is present in dry barley grains and in 5 day-old germinating barley grains. Figure 9.26 gives a summary of his investigation.
a) With reference to table 9.9, describe the procedure that you would carry out to test if a sample of unknown liquid contains
   (i)  starch
   (ii) simple sugar. (4)

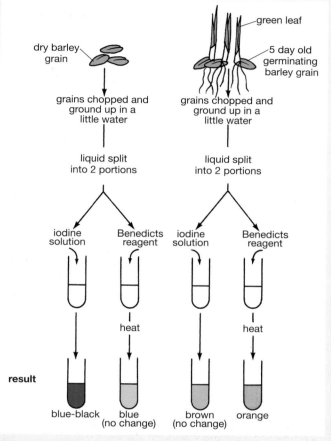

**Figure 9.26**

b) Which type of food did the student find to be present in the dry barley grains? (1)
c) (i)  Which type of food did the student find to be present in the 5-day-old germinating barley grains?
   (ii) Suggest TWO sources of this food.
   (iii) What is this type of food used for by the 5-day-old germinating seedlings? (4)
d) Table 9.10 on page 230 shows the percentages of different classes of food present in a type of nut. Draw a pie chart of this information. (2)

| food being tested for | name of chemical used for test | colour of chemical at start | colour indicating presence of food | heat needed? |
|---|---|---|---|---|
| simple sugar | Benedict's reagent | blue | orange | yes |
| starch | iodine solution | brown | blue–black | no |

**Table 9.9**

| class of food | percentage |
|---|---|
| protein | 25 |
| fat | 50 |
| carbohydrate | 10 |
| others | 15 |

**Table 9.10**

4 Figure 9.27 shows an experiment set up to find out if mustard seeds need oxygen to germinate.

a) Give THREE changes that need to be made to the set-up to make it a fair experiment. (3)

b) Explain the importance of the air-tight seal. (1)

c) Name TWO *other* factors that seeds need to germinate. (2)

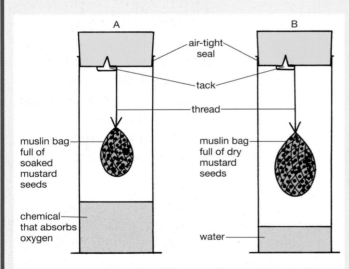

**Figure 9.27**

5 Read the following passage and answer the questions that follow it.

Banksia (see figure 9.28) is a flowering plant of the Australian bush. Following pollination and seed formation, the flowers die. However instead of dropping off, the seeds remain firmly attached to the parent plant for years waiting for the arrival of an agent which is disastrous to most living things – a bush fire.

The seeds of *Banksia* not only survive the fire but actually need it to open their seed coats. A seed coat is made up of two woody halves hinged together. It only opens when it has been thoroughly dried out by fire.

Once the fire has passed, the paper-thin, winged seeds need wind to spread them among the ashes where they will germinate when the rains arrive.

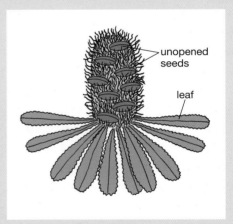

**Figure 9.28**

a) Choose the most suitable title for the passage from the following:

A   Seed dispersal in Australian flowering plants

B   The role of fire in *Banksia's* life cycle

C   Flammable agents of seed dispersal in flowers

D   *Banksia's* dependence on fire for pollination (1)

b) Choose TWO phrases from the passage that show that *Banksia* is very different from other flowering plants. (2)

c) Which condition needed by seeds for germination is given in the last sentence? (1)

**6** The following information gives some of the reasons why seeds, freshly shed into the surrounding environment by their parent plant fail to germinate.

Some seeds (e.g. sycamore) do not germinate until the seeds have been exposed to a period of low temperature just above freezing. Some seeds (e.g. orchid) remain dormant for a period of time because their embryos are still immature at the time the seeds are shed. Some seeds (e.g. sweet pea) will not germinate until their seed coat is broken open. Some seeds (e.g. tomato) are prevented from germinating by chemicals present in the surrounding fruit.

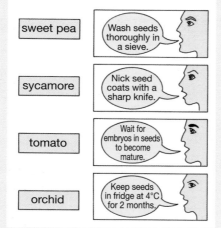

**Figure 9.29**

Copy figure 9.29 and then complete it by drawing arrows to join up each type of seed with the correct instruction. (4)

**7** The seeds of a certain cone-bearing tree are known to stay dormant for a long period after they have been shed by the parent plant. Fred and Suzi tried to break the dormancy by keeping some of the seeds at different temperatures for different lengths of time and then returning them to room temperature (20°C). Table 9.11 gives a summary of their results.

a) For how long did the seeds need to be kept at 5°C to break their dormancy? (1)

b) Which temperature broke the seed dormancy most rapidly? (1)

c) By what means could the investigation be improved to find out the very best temperature for breaking dormancy in this type of seed? (2)

d) When they looked at their results, Fred said that the seeds at −1°C failed to germinate because they had died. Suzi said they were still alive but dormant. Describe the experiment that could be carried out to see who was correct. (3)

| temperature at which seeds were kept (°C) | length of exposure time (weeks) | | | | |
|---|---|---|---|---|---|
| | 0 | 2 | 4 | 6 | 8 |
| −1 | ⬭ | ⬭ | ⬭ | ⬭ | ⬭ |
| 1 | ⬭ | ⬭ | ⬭ | 🌱 | 🌱 |
| 5 | ⬭ | ⬭ | ⬭ | ⬭ | 🌱 |
| 20 | ⬭ | ⬭ | ⬭ | ⬭ | ⬭ |

**Key** ⬭ = Seed stays dormant 🌱 = Seed begins to germinate

**Table 9.11**

| time from sowing seeds (days) | dry mass of 100 seeds germinating in darkness (g) | dry mass of 100 seeds germinating in light (g) |
| --- | --- | --- |
| 0 | 24 | 24 |
| 5 | 19 | 18 |
| 10 | 12 | 13 |
| 15 | 8 | 8 |
| 20 | 5 | 14 |
| 25 | 3 | 51 |

**Table 9.12**

**8** In an experiment to investigate the effect of light on change in dry mass of germinating pea seeds, 12 pots of 100 seeds were set up. Six pots were kept in darkness and six were kept in light. Each week the seeds from one pot in dark and one pot in light were dried to constant mass in an oven. Table 9.12 shows the results.

a) Draw a line graph of the data with 'time from sowing seeds' on the x-axis. Plot the points and carefully join them up. Use a different colour for each group of seeds. (4)

b) (i)  For how many days from the start did the seeds germinating in darkness continue to lose weight?

(ii) By which day had the seedlings germinating in darkness used up 50% of their original food store?

(iii) Extend your x-axis and predict the day on which the seeds in darkness will run out of food. (3)

c) (i)  Identify the last day on which seeds germinating in light decreased in mass.

(ii) What changes had taken place to allow them to gain mass after this day? (2)

d) By how many times was the dry mass of the seeds in light greater than that of the seeds in darkness on day 25? (1)

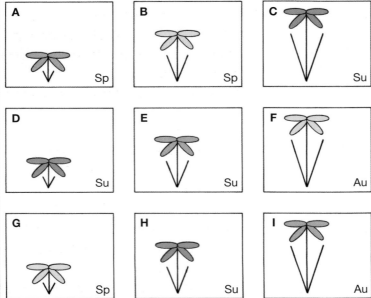

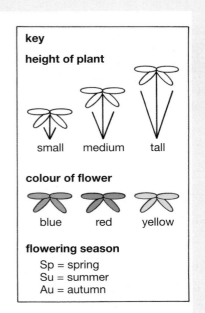

**Figure 9.31**

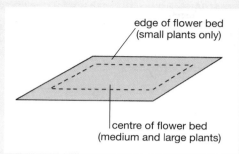

**Figure 9.30**

**9** A gardener is planning to sow seeds in the flower bed shown in figure 9.30. Figure 9.31 on page 232 gives information about the plants that develop from nine types of seed (A–I).

a) How many of the seed types will grow into plants of a height suitable for use at the edges of the flower bed? (1)

b) How many of the seed types will grow into plants of a height suitable for use at the centre of the flower bed? (1)

c) Which seed type should he sow at the edges of the flower bed to give plants with red flowers in summer? (1)

d) Which seed type should the gardener sow at the centre of the flower bed to give blue flowers in autumn? (1)

e) Give TWO differences between the plants that develop from seed types F and G. (2)

| crop | average number of seeds in 1 g |
|------|-------------------------------|
| cabbage | 220 |
| carrot | 660 |
| leek | 310 |
| lettuce | 530 |
| radish | 170 |

**Table 9.13**

**10** Table 9.13 shows the average number of seeds present in one gram of each of five different types of crop plant.

a) Which seed type is the lightest? (1)

b) Draw a bar chart of the data. (3)

c) How many more seeds are present in one gram of lettuce seeds than in one gram of leek seeds? (1)

d) By how many times is the number of seeds present in a gram of carrot seeds greater than the number of seeds in a gram of cabbage seeds? (1)

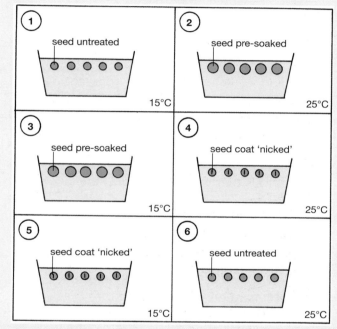

**Figure 9.32**

**11** Figure 9.32 shows six experiments set up to investigate the effects of pre-soaking, 'nicking' and temperature on germination of a type of seed.

a) Which TWO experiments should be compared to investigate the effect on germination of

(i) pre-soaking the seeds and then keeping them at 25°C?

(ii) 'nicking' the seed coats and then keeping the seeds at 15°C? (2)

b) Which TWO experiments should be compared to investigate the effect of temperature on germination of seeds that have been (i) pre-soaked? (ii) 'nicked'? (2)

| seed type | chemical in layer of pellet | pest or fungal disease being targeted |
|---|---|---|
| carrot | fungicide W | damping off |
| cauliflower | fungicide W | downy mildew |
| cabbage | insecticide X | stem beetle |
| cauliflower | insecticide X | stem beetle |
| bulb onion | fungicide Y | damping off |
| cabbage | fungicide Y | damping off |
| carrot | fungicide Y | damping off |
| cauliflower | fungicide Y | damping off |
| bulb onion | insecticide Z | seed fly |
| carrot | insecticide Z | carrot fly |

**Table 9.14**

**12** Table 9.14 gives information about chemicals that may be added to the layers when seeds are pelleted.

a) Name THREE insect pests given in the table. (3)

b) Is *damping off* a pest or a disease? (1)

c) Which TWO chemicals can be put in pellets to prevent *damping off*? (2)

d) How many different pests or diseases that attack carrot plants are given in the table? (2)

e) Name THREE different pests or diseases suffered by cauliflower plants. (3)

f) How many different plant types are given as sufferers of *damping off*? (1)

## What you should know

| | | |
|---|---|---|
| advantage | germination | pesticides |
| Albustix | green | photosynthesise |
| blue-black | hand | pre-soaking |
| Clinistix | iodine | purple |
| coat | light | root |
| cycle | longer | sand |
| dormant | loss | slitting |
| embryo | low | spaced |
| gain | oxygen | spring |
| generation | pelleted | temperature |

**Table 9.15** *Word bank for chapter 9*

1 A seed contains an embryo _____ and an _____ shoot and their food store. These are surrounded by a protective seed _____.

2 Different types of food are found in the food store of seeds. The presence of starch can be demonstrated using _____ solution. If positive, the result is a _____ colour. The presence of glucose can be demonstrated using a _____ strip. If positive, the result is a _____ colour. The presence of protein can be demonstrated using an _____ strip. If positive, the result is a _____ colour.

3 During _____, an embryo plant uses the reserves of food in the seed to grow into a plant with green leaves.

4 Seeds need water, _____ and a suitable _____ (e.g. 20°C) to germinate.

5 A plant's life _____ is the series of changes that it passes through from a certain stage in its development (e.g. germinating seed) until it reaches the same stage in the next _____.

6 Seeds which remain inactive even when given the conditions needed to germinate, are described as being _____. One type of dormancy can be broken by exposing the seeds to _____ temperature (e.g. 4°C) for a few months. Under natural conditions this form of dormancy is of _____ to the plant because germination of the seeds is delayed until _____ when the warmer soil temperature encourages growth.

7 During photosynthesis, green plants use _____ energy to produce food needed for growth. Embryo plants inside germinating seeds cannot _____ and need to use the seed's food reserves for growth. Therefore germinating seeds show an overall _____ in mass. Photosynthesising seedlings are able to produce the food they need for growth and therefore show an overall _____ in mass.

8 To sow seeds means to put them in a place which provides the conditions needed for germination. Seeds need to be _____ out during sowing to prevent overcrowding of seedlings following germination. Large seeds are sown individually by _____; fine seeds are mixed with silver _____ before sowing.

9 Seeds that have been enclosed in a ball of clay are said to have been _____. The pellets make the seeds smooth and easy to space out during planting. A pellet is made up of layers which may contain useful chemicals such as _____ and/or nutrients. Pelleted seeds take _____ to germinate than unpelleted seeds.

10 To pre-germinate (chit) a seed means to make it start to germinate before it has been planted. In some seeds this can be done by _____ them. Seeds with hard coats can be made to germinate more quickly by _____ them open or 'nicking' their seed coats with a knife before sowing.

## 10  Vegetative propagation

### Propagation

**Propagation** is the process by which a grower increases his/her supply of a type of plant. Plants can be propagated using **seeds** (see chapter 9). Since the seeds are formed **sexually** from male and female sex cells, the plants that develop are *not* all identical.

### Vegetative propagation

Plants can also be propagated **vegetatively**. This method makes use of various structures such as bulbs and cuttings that are formed by **non-sexual** (asexual) means. All the plants that develop from them are **identical** to the parent plant. This is very useful to the grower when s/he wants to increase the supply of an attractive and/or valuable type of plant. The grower can depend on all the offspring being exactly the same. In addition, vegetative propagation is normally faster than propagation by seed.

### Types of vegetative propagation

Vegetative propagation may be natural or artificial. During **natural vegetative propagation**, the supply of the plant is increased by making use of structures (e.g. bulbs) formed naturally by the plant. During **artificial propagation**, the supply of the plant is increased using structures that have been cut off or developed from the plant in some way that the plant is unable to do naturally. Figure 10.1 gives a summary of the means of plant propagation.

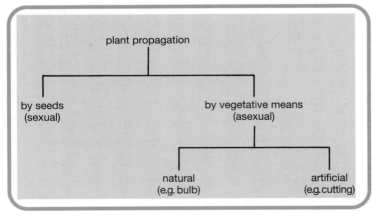

**Figure 10.1**  *Means of plant propagation*

## Use of natural plant propagation structures

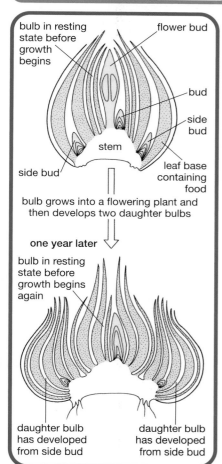

bulb in resting state before growth begins

flower bud

bud

side bud

stem

side bud

leaf base containing food

bulb grows into a flowering plant and then develops two daughter bulbs

one year later

bulb in resting state before growth begins again

daughter bulb has developed from side bud

daughter bulb has developed from side bud

**Figure 10.2** *Vegetative propagation in a daffodil bulb*

### Food storage organs

#### Bulbs

A **bulb** is a natural plant propagation structure. It is made up of several **leaf bases** swollen with stored food and attached to a short thick stem.

Side buds are found amongst the leaf bases. These develop into **daughter bulbs**. Figure 10.2 shows a section through a Daffodil bulb before growth and then shows the same bulb a year later.

### Activity

## Investigating the types of food stored in bulbs

### You need

- half of a daffodil bulb cut lengthways
- half of an onion bulb cut lengthways
- forceps
- pestle (grinder) and mortar (bowl)
- dropping bottle of iodine solution
- Clinistix strips
- Albustix strips
- ceramic tile
- hand lens
- measuring cylinder (10 cm³ plastic)
- two test tubes and test tube stand
- silver sand

**Figure 10.3** *Daffodil bulbs*

### What to do

1. Before starting the experiment, read all of the following steps and then construct a results table for the three food tests.
2. Examine the cut surface of the daffodil bulb carefully and identify a) the short stem and b) the swollen leaf bases. (Also see figure 10.3.)
3. With the aid of the hand lens, look for side buds.
4. Using the forceps, pick out a piece of leaf base (about the size of a 2p coin) and test it for starch using three drops of iodine solution (positive result = blue–black). Enter the result in your table.
5. Pick out a second piece of leaf base and grind it up in the mortar with 5 cm³ of water and a little silver sand.
6. Divide the liquid between the two test tubes.

7 Test one sample for glucose using Clinistix strip (positive result = purple) and the other sample for protein using an Albustix strip (positive result = green). Enter the results in your table.

8 Clean the pestle, mortar and test tubes as instructed.

9 Repeat steps 2–8 using the half of onion bulb. (Also see figure 10.4.)

10 Draw conclusions from your results.

**Figure 10.4**   *Onion bulbs*

## Activity

### Planting and growing a daffodil bulb

**You need**

- label
- crocks
- plant pot (e.g. 100 mm top diameter)
- drainage saucer
- plastic bag containing a potful of moist bulb fibre (or compost rich in peat)
- daffodil bulb
- measuring cylinder (10 cm³ plastic)

**What to do**

1 Label your pot as instructed.

2 Position crocks in the bottom of the pot to cover the drainage holes (see figure 10.5).

3 Place a layer of bulb fibre (e.g. 30 mm) in the bottom of the plant pot and firm it gently with your fingers.

4 Stand the bulb, roots down, in the bulb fibre with the top of the bulb level with the rim of the pot.

5 Add more bulb fibre around the bulb and firm it.

6 Continue until the fibre is about 10 mm from the top of the pot and the bulb is just sticking out of the compost.

7 Stand the pot in its drainage saucer and add a small volume of water (e.g. 10 cm³) using the measuring cylinder.

8 Allow any excess water to drain into the saucer and then empty it.

9 Place the pot and saucer in a cold, dark place for about 8 weeks.

10 Bring the pot back into the light when the shoot is about 20 mm long.

11 Keep the fibre moist.

12 Inspect your bulb regularly over the next few months as it flowers and then produces more daughter bulbs by vegetative propagation.

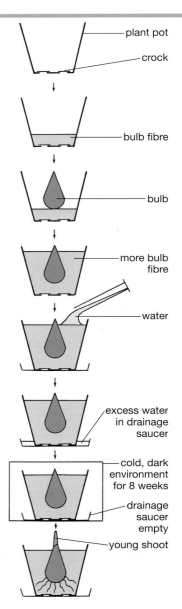

**Figure 10.5**   *Planting a daffodil bulb*

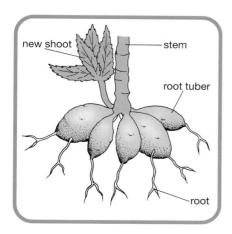

**Figure 10.7** *Root tubers in dahlia*

### Planting bulbs in a garden

Spring flowering bulbs are usually planted in autumn. The planting depth in the garden depends on the size of the bulb. Normally bulbs are planted 2–3 times deeper than their length and 2–3 bulb widths apart (see figure 10.6).

### Propagation of bulbs

Many types of bulbs (e.g. daffodil) increase in number naturally by forming daughter bulbs around the original parent bulb (see figure 10.2). These daughters can be separated from the parent and planted in new sites.

### Tubers

A **tuber** is a swollen region of a stem or root which is a food storage organ. During summer, a dahlia plant forms many **root tubers** (see figure 10.7). Each of these tubers is capable of growing into one or more Dahlia plants if it is separated from the others and planted in a new site the next spring.

During summer, a potato plant forms many **stem tubers** at the ends of underground stems (see figure 10.8). Each tuber has 'eyes' which are really tiny buds. When the tuber is planted in a new site, its eyes develop into new shoots. These get an early start by using food stored in the tuber. Each tuber grows into a potato plant with one or more underground shoots that in turn produce another generation of stem tubers by vegetative propagation.

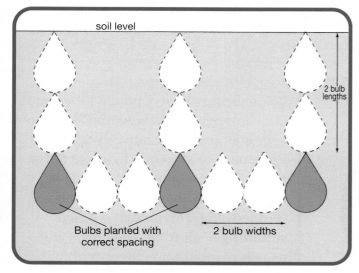

**Figure 10.6** *Depth and spacing of bulbs*

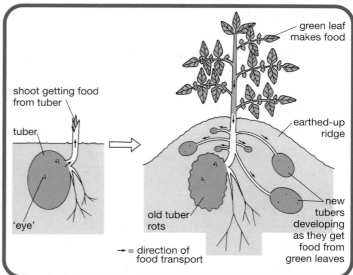

**Figure 10.8** *Stem tuber formation in a potato plant*

## Investigating the types of food stored in a potato tuber

### You need
- half of a washed potato tuber with 'eyes'
- spatula
- pestle and mortar
- dropping bottle of iodine solution
- Clinistix strips
- Albustix strips
- ceramic tile
- hand lens
- measuring cylinder (10 cm³ plastic)
- two test tubes and test tube rack
- silver sand

### What to do
1 Before starting the experiment, read all of the following steps and then construct a results table for the three food tests.
2 Examine the outer surface of the potato and find an 'eye'.
3 Carefully examine the eye using a hand lens.
4 Using the spatula, scoop out a small sample of potato tissue on to the tile and test it for starch using three drops of iodine solution (positive result = blue–black). Enter the result in your table.
5 Scoop out a second sample and grind it up in the mortar with 5 cm³ water and a little silver sand.
6 Divide the liquid between the two test tubes.
7 Test one sample for glucose using a Clinistix strip (positive result = purple) and the other sample for protein using an Albustix strip (positive result = green). Enter the results in your table.
8 Draw a conclusion from your results.

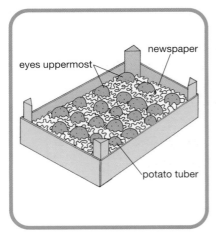

**Figure 10.9** *Potatoes left to sprout*

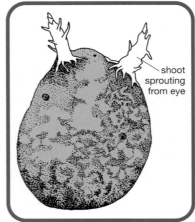

**Figure 10.10** *A chitted potato*

Preparing potato tubers for planting

Before being planted, potato tubers are often arranged in an open tray with the ends containing the most '**eyes**' uppermost as shown in figure 10.9. The tray is kept in a cool room for about 6 weeks to allow the buds in the eyes to start growing. This growth is called **sprouting** or **chitting**. The tubers are usually planted when the sprouts are about 20 mm long (see figure 10.10)

This procedure cuts down growing time because chitted tubers produce potato plants and the next generation of potato tubers much more quickly than non-chitted tubers.

## Planting and growing a potato tuber

*You need*
- label
- potato tuber (preferably chitted)
- plant pot (e.g. top diameter 150 mm or larger)
- drainage saucer
- crocks
- moist, loamless multipurpose compost (rich in humus, free of lime)
- beaker (100 cm$^3$ plastic)
- cane and wire

*What to do*

1 Label your pot as instructed.
2 Use crocks to cover the holes at the bottom of the pot.
3 Half-fill the pot with compost.
4 Firm the compost gently using the plastic beaker.
5 Plant the potato tuber with its sprouting shoots upwards.
6 Fill the pot with compost to 10 mm from the top and then firm gently.
7 Water the compost thoroughly and allow the pot to drain into the sink.
8 Sit the pot in its drainage saucer in a cool room (e.g. 16–18°C).
9 Keep the compost moist with frequent watering.
10 Inspect your potato plant regularly over the next few months as the shoots and green leaves develop.
11 Support the mature plant (which may reach a height of 600 mm) with a cane.
12 Dig up the compost to reveal your crop of new potato tubers formed by vegetative propagation (see figure 10.11).

**Figure 10.11** *Potato tubers*

## Use of attached offspring

### Plantlet

A **plantlet** is a tiny version of a plant still attached somewhere to its parent. Some plants such as the Mexican hat plant (*Bryophyllum*) develop plantlets on the edges of mature leaves (see figure 10.12).

**Figure 10.12** *Attached plantlets on a Mexican hat plant*

**Figure 10.13**
*Attached plantlet on a piggyback plant*

In other plants, such as a piggyback plant (see figure 10.13), a plantlet develops at the base of each mature leaf where it meets the leaf stalk.

## Activity

### Growing a plant from a plantlet

You need

- label
- plant pot (e.g. 80 mm top diameter)
- drainage saucer
- crocks
- moist compost (rooting or multipurpose)
- plastic forceps
- scissors
- beaker (100 cm³ plastic)
- access to mature Mexican hat plant or piggyback plant with plantlets

What to do (also see figure 10.14)

1 Label your pot as instructed.
2 Use crocks to cover the holes at the bottom of your pot.
3 Fill the pot with compost.
4 Using the plastic beaker, firm the compost down to 20 mm below the rim of the pot.
5 Water the compost and allow the excess water to drain into the sink.
6 Sit the pot in its drainage saucer.

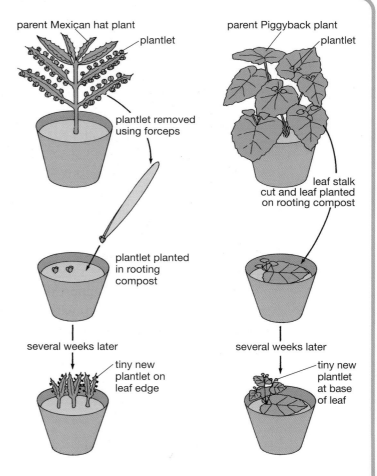

**Figure 10.14** *Growing plants from plantlets*

7 For the Mexican Hat plant, follow steps 8–10 below. For the piggyback plant, follow steps 11–12 below.

8 Go to the Mexican hat plant and, using forceps, remove one of the plantlets holding it by a leaf.

9 Place the plantlet, roots down, on the moist compost in your pot.

10 Repeat the procedure for two more plantlets.

11 Go to the parent Piggyback plant and remove a mature leaf bearing a plantlet by cutting through the leaf stalk about 20 mm from the leaf base.

12 Push the stump of the leaf stalk into the moist compost in your pot and gently press the leaf down so that the base bearing the plantlet is touching the moist compost.

13 Do not give your plant more water until the plantlet has rooted and the compost has started to dry out. Then water as required.

14 Inspect your plantlet regularly over the weeks that follow and watch it develop into an adult plant bearing the next generation of plantlets formed by vegetative propagation.

### Runner

Some types of plant produce plantlets at the ends of 'horizontal' stems called **runners**. Figure 10.15 shows a spider plant (*Chlorophytum*) with runners bearing plantlets. Figure 10.16 shows a mother-of-thousands plant with runners bearing plantlets.

**Figure 10.15**  *Runners bearing plantlets in a spider plant*

**Figure 10.16**  *Runners bearing plantlets in a mother-of-thousands plant*

## Growing a plantlet formed at the end of a runner

**You need**
- label
- plant pot (e.g. 60 mm top diameter)
- drainage saucer
- crocks
- moist compost (rooting or multipurpose)
- scissors
- U-shaped length of nicrome wire
- beaker (100 cm³ plastic)
- access to mature spider plant with runners bearing plantlets
- spatula

**What to do**
1 Label your pot as instructed.
2 Cover the drainage holes with crocks.
3 Fill the pot with compost.
4 Using the plastic beaker, firm the compost down to 20 mm below the rim of the pot.

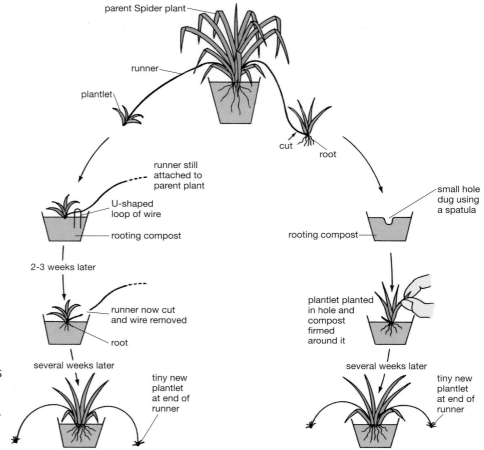

**Figure 10.17** *Growing plantlets formed at the ends of runners*

5 Water the compost and allow the excess water to drain into the sink.
6 Sit the pot in its drainage saucer.
7 Go to the parent spider plant and inspect the plantlet that you are going to use. If it does not have roots, follow steps 8–9 below. If it already has roots, follow steps 10–12 below. In both cases see figure 10.17.
8 Using the U-shaped length of wire, peg down the runner to hold the plantlet in contact with the moist compost in your pot.
9 Around 2–3 weeks later, when the plantlet has rooted, cut the runner with scissors.
10 Using the spatula, make a small hole in the compost in your pot.
11 Cut the runner 5 mm from the plantlet using scissors.
12 Gently lift the plantlet by its leaves, plant it in the hole and firm the compost around it.
13 Do not give your pot more water until the compost is beginning to dry out. Then water as required.
14 Inspect your plantlet regularly over the weeks that follow. Watch it develop into an adult spider plant with runners bearing the next generation of plantlets formed by vegetative propagation.

**Figure 10.18** *Offsets*

## Offsets

An **offset** is a tiny plantlet that develops as a side shoot at the base of the parent (see figure 10.18). Sometimes several different offsets are produced and form a clump at the base of the parent plant.

A plant with offsets can be propagated by **division**. This means separating the offsets from the parent, normally using a sharp knife, and planting them in new locations.

## Activity

### Separating and growing an offset

**You need**
- label
- crocks
- plant pot (e.g. 60 mm top diameter)
- drainage saucer
- moist rooting compost (see page 268)
- grit
- spatula
- beaker (100 cm³ plastic)

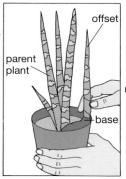

offset's base uncovered

offset cut away from parent plant

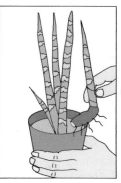

offset and attached roots carefully lifted

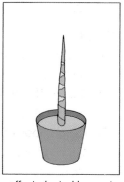

offset planted in a pot of rooting compost

**Figure 10.19** *Separating and planting an offset of a mother-in-law's tongue plant*

- supply of offsets cut from parent plant of aloe or mother-in-law's tongue (to be prepared by teacher)

**What to do**
1. Label your plant pot as instructed.
2. Cover the drainage holes with crocks.
3. Fill the pot with compost.
4. Using the plastic beaker, firm the compost down to 20 mm below the rim of the pot.
5. Dig a small hole in the centre of the compost.
6. Watch as the teacher prepares the offsets for use. (This is done by scraping away the surface compost beside the offsets in the parent plant's pot. Each stem offset is then cut off as near to the parent stem as possible. Care should be taken to preserve any roots that are attached to an offset. Also see figure 10.19.)

7 Holding your offset by its leaves, plant its base in the compost.

8 Firm the soil around it.

9 Top dress the compost by sprinkling a very thin layer of grit over it.

10 Place the pot in a drainage saucer and water it after 3–4 days.

11 Inspect your offset regularly over the months that follow. Watch it develop into an adult plant with a new generation of offsets developing at the base of the stem by vegetative propagation.

## Testing your knowledge

1 What is the difference between *seed* propagation and *vegetative* propagation? (2)

2 Redraw figure 10.20 and use lines to correctly pair each plant propagation structure with the type of plant that makes it. (5)

3 Describe how you would carry out tests to find out if a bulb's storage leaves contained a) starch and b) glucose. (4)

4 Briefly describe how you would grow a plant from a plantlet of a piggyback plant. (Assume that you already have your pot of moist compost prepared.) (4)

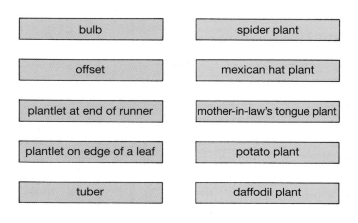

| bulb | spider plant |
| offset | mexican hat plant |
| plantlet at end of runner | mother-in-law's tongue plant |
| plantlet on edge of a leaf | potato plant |
| tuber | daffodil plant |

**Figure 10.20**

5 Decide whether each of the following statements is true or false and use T or F to indicate your choice. Where a statement is false, give the word that should have been used in place of the word(s) in **bold** print. (5)

a) Pots containing freshly planted bulbs should be left in a **warm**, **bright** place for several weeks.

b) Onion plants increase their number vegetatively by producing **tubers**.

c) A chitted potato tuber is one whose eyes have sprouted into **shoots** before being planted.

d) Dahlia plants increase their numbers vegetatively by producing **root** tubers.

e) Plantlets and offsets are examples of **detached** offspring.

# Artificial propagation

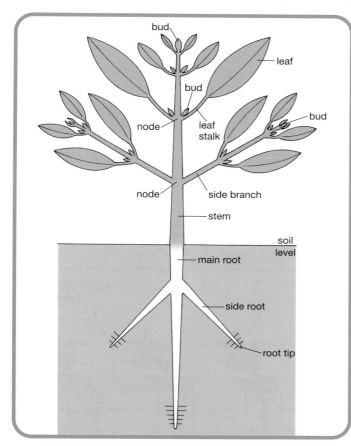

**Figure 10.21**  *A young plant's growing points*

**Artificial propagation** means increasing the supply of a type of plant using some method that the plant itself would be unable to carry out.

## Growing points

Figure 10.21 shows a simple diagram of a young plant. Under natural conditions, growth takes place at several points on a plant such as the **buds** and **root tips**.

### Node

A **node** (see figure 10.22) is a region of stem from which one or more leaves or side branches grow. A node is therefore a point of growth.

### Wounding

When a plant is **wounded**, the cells at the site of the wound start to divide and grow. Figure 10.23 shows what can happen at several sites of wounding when a plant is cut up or has a side branch ripped off.

**Figure 10.22**  *Nodes*

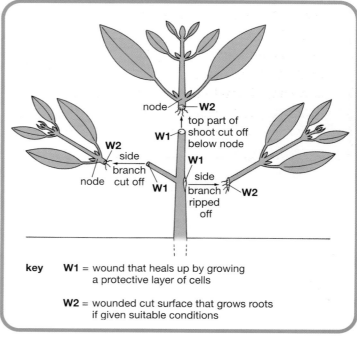

key     W1 = wound that heals up by growing
             a protective layer of cells

         W2 = wounded cut surface that grows roots
             if given suitable conditions

**Figure 10.23**  *Effects of wounding a plant*

**Figure 10.24** *"Cut them out and transplant them. That's my advice."*

## Cuttings

When people take plant **cuttings**, they are bringing about artificial propagation. This process makes use of the ability of plants to grow new cells at wounds and replace missing parts (see figure 10.24). In many plants, the growth of roots directly from the cut surface of a stem occurs easily when the cut end is planted in moist rooting compost or left in water for a week or two (see figure 10.25).

## Rooting powder

In some plants, the cut stem does not develop roots easily. In these cases, the grower often dips the cut end into **rooting powder** before planting (see figure 10.26). This contains a chemical that encourages the cells at the cut surface to divide and develop into roots (see figures 10.27 and 10.28).

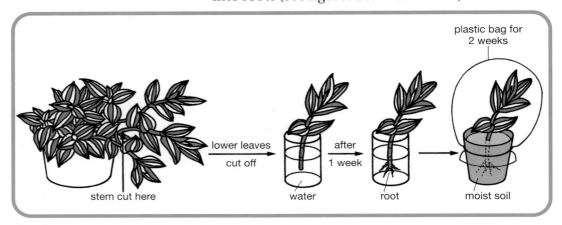

**Figure 10.25** *Taking a stem cutting of a wandering sailor plant*

**Figure 10.26** *Using rooting powder*

## Reducing water loss

A newly planted stem cutting does not have roots that it can use to absorb water. However its leaves continue to lose water (as water vapour) through tiny holes on their surfaces. The grower has to find ways of reducing this water loss until the cutting has developed roots. Two ways of doing this are as follows.

### Reducing the leaf surface area of the cutting

Some of the lower leaves are trimmed off the cutting before it is planted. At the same time the upper leaves and buds are left in place to allow photosynthesis and growth to occur.

**Figure 10.27** *Roots from a cut stem – early stage*

A **propagator** is a piece of equipment that provides a humid atmosphere for raising young plants such as stem cuttings. This can be a very simple structure such as a **plastic bag** or a **cloche** made from the base of a plastic bottle (see figure 10.29).

It is important that the plastic bag or cloche does not touch the plant and soak it with condensation since this can lead to leaf rot. A framework of canes or wire can be used to keep a plastic bag up and away from the cutting.

Pots with their enclosed cuttings should be kept in a shaded place which provides some light for photosynthesis while at the same time preventing high temperatures building up.

A propagator can be a more complex piece of equipment (see figure 10.30) which may have a **heating** system and a **misting** system. The heating system (normally a soil-warming cable buried in sand) encourages root growth. The misting system provides a moist environment which stops the leaves drying out and wilting. These systems are often connected to sensors and switch on and off automatically as required.

**Figure 10.28** *Roots from a cut stem – later stage*

plastic bag

base of plastic bottle

**Figure 10.29** *Simple propagators*

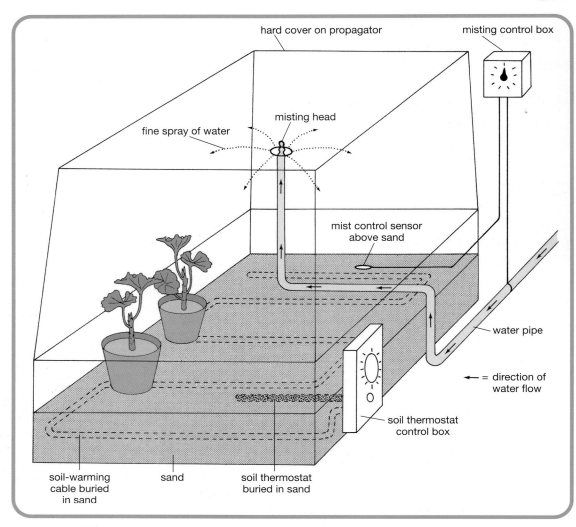

Labels in figure:
- hard cover on propagator
- misting control box
- misting head
- fine spray of water
- mist control sensor above sand
- water pipe
- = direction of water flow
- soil thermostat control box
- soil-warming cable buried in sand
- sand
- soil thermostat buried in sand

**Figure 10.30** *Mist propagator*

### Activity

## Taking a stem cutting (assessment)

You need
- label
- plant pot (e.g. top diameter 80 mm)
- crocks
- drainage saucer
- beaker (100 cm³ plastic)
- moist rooting compost
- secateurs (see figure 10.31)
- rooting powder
- dibber or pencil
- access to parent plant (e.g. geranium or fuchsia)
- polythene bag and elastic band or access to propagator

**Figure 10.31** *Secateurs*

## What to do

1 Label your pot as instructed.

2 Cover the holes in your pot with crocks.

3 Fill the pot with compost and firm it down to 10 mm below the rim of the pot using the plastic beaker.

4 Using a dibber (or pencil) make a hole (about 50 mm deep) at the centre of the compost.

5 Prepare the stem cutting as shown in figure 10.32.

6 Unlike geranium in figure 10.32, some types of plant cutting do not form roots easily and need to have rooting powder applied to them. If instructed by the teacher, dip the end of your cutting into rooting powder. Lift it out carefully and tap the stem to knock off any excess powder.

7 Plant the cutting in the hole in the compost so that approximately the same length of stem is underground as above soil level. Firm the compost around the stem.

8 Water the compost and allow the excess water to drain down the sink.

9 Place the pot in its draining saucer.

10 Enclose your cutting in a polythene bag or place it in the plant propagator once your teacher has assessed your work.

11 After a few weeks, when your cutting has rooted, remove the polythene bag or take the pot out of the propagator and let your plant grow in normal conditions.

12 When it reaches maturity, repeat the process of artificial propagation by taking stem cuttings from your plant.

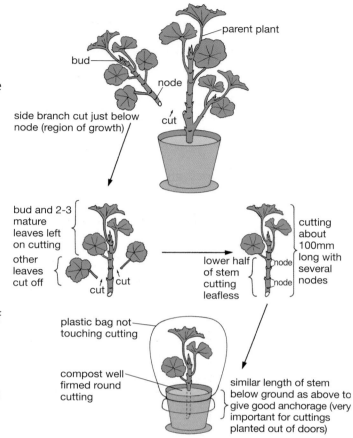

**Figure 10.32** *Preparing and planting a stem cutting*

## Activity

# Taking a leaf cutting of African violet plant

## You need

- label
- plant pot (e.g. top diameter of 80 mm)
- crocks
- drainage saucer
- cloche (base cut from 2 litre plastic bottle)
- African violet compost
- dibber (or pencil)
- leaf (with leaf stalk about 50 mm in length) from parent African violet plant
- beaker (100 cm³ plastic)
- access to plant propagator at 24°C

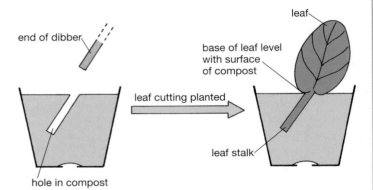

**Figure 10.33** *Planting a leaf cutting*

## What to do

1 Label your pot as instructed.
2 Cover the pot's drainage holes with crocks.
3 Fill the pot with compost.
4 Use the plastic beaker to firm the compost to 20 mm below the pot's rim.
5 Add about 100 cm³ water to moisten the compost allowing any excess to drain into the sink.
6 Using a dibber (or pencil) make a hole in the compost about 50 mm deep and at an angle (see figure 10.33).
7 Plant the leaf stalk in the hole so that the leaf base is level with the surface of the compost (see figure 10.33).
8 Carefully firm the compost around the base of the leaf without covering up any of the leaf.
9 Add a cloche to prevent the leaf cutting from drying out.
10 Sit the pot in the drainage saucer and place it in the propagator to provide warmth for root growth.
11 Over the weeks that follow, check your leaf cutting and water it from below only if necessary.
12 After several weeks, when your leaf cutting has produced one or more plantlets (see figure 10.34), remove the cloche. Take the pot out of the propagator and keep it under normal conditions.
13 When your plantlets have developed into mature plants, repeat the process of artificial propagation by taking leaf cuttings from them.

**Figure 10.34** *Plantlet from a leaf cutting of African violet*

### Layering

**Layering** is a method of artificial propagation by which a normal stem (not a runner) is made to root while *still attached* to the parent plant. A vigorous, non-flowering side shoot with several nodes is chosen.

A small cut is made into the stem just below a node (see figure 10.35) and rooting powder is applied to this 'wound'. The stem is pegged down into the soil at the wound to allow roots to form at the node. The end of the shoot is tied to a cane to keep it upright. Once roots have developed and the shoot is showing new growth, the pegged-down stem is cut to separate it from the parent plant. A new plant has been produced by artificial propagation.

Some trailing plants (e.g. wandering sailor) readily develop roots on the nodes of side branches without being wounded or having rooting powder applied. Contact with damp compost or soil is enough to bring about layering of their side branches.

Layering is of advantage to the grower because it produces large plants that already have vigorous, healthy shoots likely to survive without being 'nursed' through their early growth in a propagator. The supply of many types of woody plant can be increased by layering but it is most commonly used for plants such as rhododendron that are difficult to grow from cuttings.

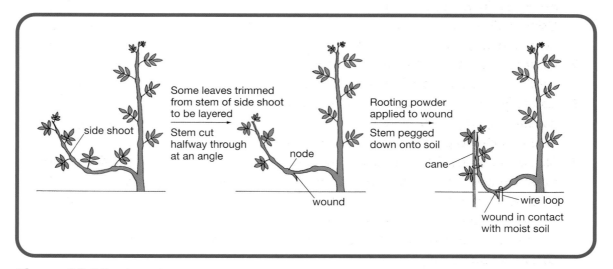

**Figure 10.35** *Layering*

## Propagating a plant by layering

### You need

- label
- plant pot (e.g. top diameter of 80 mm)
- drainage saucer
- crocks
- U-shaped length of wire
- cane
- twine
- potted plant with long, trailing stem (e.g. ivy)
- moist rooting compost
- beaker (100 cm³ plastic)
- scissors
- rooting powder

### What to do

1  Label your pot as instructed.
2  Cover the pot's drainage holes with crocks.
3  Fill the pot with compost.
4  Use the plastic beaker to firm down the compost fairly hard.
5  Add 100 cm³ of water and refirm the compost.
6  Sit the pot on its drainage saucer.
7  Select a healthy, vigorous, non-flowering shoot on the parent plant (see figure 10.36).
8  Using scissors, trim the leaves from the centre section of the stem giving a cleared length of about 100 mm.

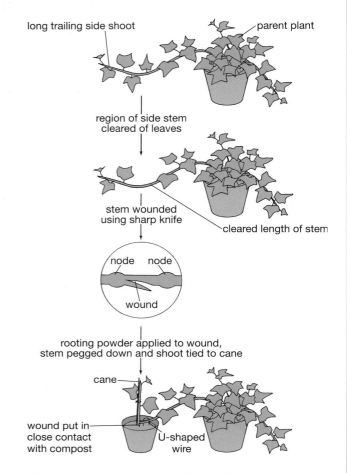

**Figure 10.36**  *Propagation of an ivy plant by layering*

9  Watch while your teacher wounds the cleared length of stem near its centre. (This is done by using a sharp knife to make an angled cut halfway through the stem just below a node.)
10  Apply rooting powder to the wound.
11  Firmly peg down the wounded region of the stem.
12  Tie the front end of the side shoot to a cane so that it stays upright.
13  After several weeks, when the wounded stem has rooted (and the shoot has shown new growth), cut the stem that connects the layered plant to the parent.

## Use of heat during propagation

Plant cuttings and seedlings often begin life in a propagator situated in a greenhouse. There are advantages and disadvantages of providing the young plants with heat at this stage in their development.

### Advantages

A propagator with a heating system such as a soil-warming cable keeps stem cuttings at a suitable temperature 24 hours a day. This encourages them to develop roots quickly. In an unheated greenhouse (or cold frame) the temperature varies widely during the day and night and the plants do not get ideal growing conditions.

A heating system in a greenhouse is of particular benefit in the winter months when it can be used to protect the young plants being propagated from damage by frost.

### Disadvantages

An increase in temperature in a plant's environment leads to an increase in rate of water loss by the plant through its leaves. If precautions are not taken, the plants being propagated will wilt and, in extreme cases, die.

If the plant's environment is kept very moist as well as very warm, an extremely humid atmosphere may build up. This provides ideal conditions for the spread of diseases such as **grey mould** (see page 289). In addition, heating systems use up **energy** and therefore can be expensive to operate, especially during the cold winter months.

### Suitable balance

A thermostatically controlled heating system (see also page 278) which includes a soil-warming cable provides 'bottom heat' to promote root growth ahead of shoot growth. If the temperature rises above a suitable pre-set level, the heating system switches off; if the temperature drops below a certain pre-set level, the heating comes on again. This system provides the advantages given above while keeping the disadvantages to a minimum.

## Testing your knowledge

**1** What is the difference between *natural* and *artificial* propagation? (2)

**2** a) What is a *node*? (1)
   b) Identify TWO other points of growth on a plant. (2)

**3** Give TWO ways in which a plant may respond to having its stem wounded. (2)

**4** a) Give TWO methods used by growers to reduce water loss by newly planted stem cuttings. (2)
   b) Some plants are difficult to propagate from cuttings. Name another method of artificial propagation that is often successful in such cases. (1)

**5** Give ONE advantage and ONE disadvantage of supplying heat to young plants during propagation. (2)

**6** Rewrite each of the following sentences, including only the correct answer at each choice given in **bold** print. (5)
   a) Taking a stem cutting involves making a cut **halfway/all the way through** the stem just **above/below** a node.
   b) Crocks are used to cover the holes in a plant pot so that **water/compost** stays in while excess **water/compost** can escape.
   c) Rooting powder is applied to cut **roots/shoots** to make them develop **roots/shoots**.
   d) Most of the mature leaves on a stem cutting are removed in order to cut down **water loss/photosynthesis** but two or three are left to allow **water loss/photosynthesis**.
   e) During layering, the wounded **side shoot/leaf stalk** is **pegged down into/inserted into a deep hole** in the moist compost.

## Applying your knowledge

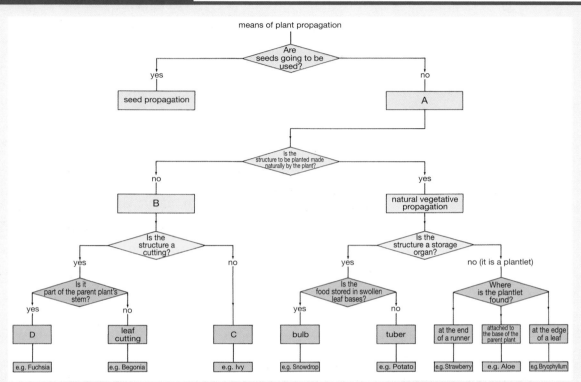

**Figure 10.37**

**1** Figure 10.37 shows a key to plant propagation.

a) Match boxes A, B, C and D on the diagram with the following terms: artificial propagation, layering, stem cutting, vegetative propagation. (2)

b) Cuttings can be taken from plant parts other than the stem. Give an example from the key and name the plant that can be propagated by this method. (2)

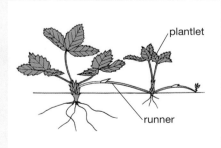

**Figure 10.38**

c) (i) Which type of plant in the diagram can be propagated naturally using storage organs that contain their stored food in swollen leaf bases?

(ii) What name is given to its type of propagation structure? (2)

d) (i) Name the type of propagation structure used by aloe and *Bryophyllum*?

(ii) Where is this structure found in each case? (3)

e) Identify the plant shown in figure 10.38 using the key. (1)

**2** Figure 10.39 shows five stages (P–T) carried out when planting and growing a hyacinth bulb. They are given in a mixed-up order.

a) Arrange stages P–U into the correct sequence. (1)

b) (i) Once the flower has died, how could you find out if the bulb had propagated itself vegetatively?

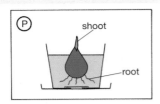

P

shoot

root

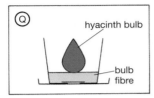

Q

hyacinth bulb

bulb
fibre

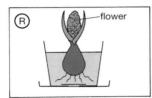

R

flower

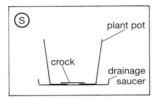

S

plant pot

crock

drainage
saucer

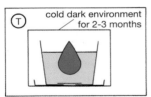

T

cold dark environment
for 2-3 months

**Figure 10.39**

   (ii) Is such propagation described as
      natural or artificial?    (2)

**3** Read the opposite passage and answer the
following questions.
  a) Why should potato tubers be chitted in
    light rather than in darkness?   (1)
  b) If you didn't know whether your tubers
    were early variety or main crop, when
    would be the best time to plant them?
    Explain why.   (2)
  c) (i)  What is meant by *earthing up* a potato
       crop?
    (ii) Give TWO reasons why earthing up is
       important.   (4)
  d) Why are plastic bags unsuitable for long-
    term storage of new potatoes?   (2)

## GROWING YOUR OWN POTATOES

The potato tubers used should first be chitted in light for about six weeks. This makes their shoots short and sturdy. If they are kept in total darkness, their shoots will be long and weak. Early varieties should be planted from March to April; main crop varieties from April to May.

Potato tubers can be planted in rows in V-shaped trenches (see figure 10.40). The trenches are filled in and the tubers left to grow. Once shoots appear and develop leaves, it is time to earth up the crop. This means pulling up soil from between the rows to make ridges (see figure 10.41). This practice stops the new tubers pushing through the soil into sunlight and becoming green. Green tubers cannot be eaten. Earthing up also kills weeds that would compete with the potato plants for nutrients in the soil.

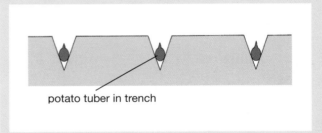

potato tuber in trench

**Figure 10.40**

The harvest of new potatoes should be stored in a cool, frost-free place. Potato tubers are alive and give out moisture. If they are stored in plastic sacks, moisture builds up and makes them rot. Brown paper sacks are best because they allow excess moisture to escape.

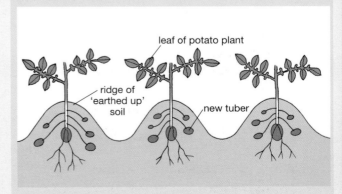

leaf of potato plant

ridge of
'earthed up'
soil

new tuber

**Figure 10.41**

| plant | type of propagation structure | hardiness score | when to plant | depth of planting (cm) | light requirements | flowering season |
|---|---|---|---|---|---|---|
| Begonia | tuber | 1 | spring | 3 | ☀ | summer |
| Crocus | corm | 4 | autumn | 6 | ☀ | spring |
| Daffodil | bulb | 5 | autumn | 12 | ☀ or ☁ | spring |
| Gladiolus | corm | 2 | spring | 10 | ☀ | summer |
| Hyacinth | bulb | 4 | autumn | 14 | ☀ | spring |
| Iris | rhizome | 4 | autumn | 7 | ☀ | summer |
| Lily | bulb | 5 | autumn | 11 | ☀ | summer |
| Snowdrop | bulb | 5 | autumn | 5 | ☁ | winter |
| Tulip | bulb | 4 | autumn | 10 | ☀ | spring |

| key to hardiness score | minimum temperature (°C) |
|---|---|
| 1 = tender | −1 |
| 2 = slightly hardy | −6 |
| 3 = fairly hardy | −12 |
| 4 = very hardy | −18 |
| 5 = extremely hardy | −29 |

key to light requirements

 = sun

 = semi-shade

**Table 10.1**

4 Table 10.1 gives information about propagating nine types of plant.

a) How many plant types in the table produce a propagation structure other than a bulb? (1)

b) How many of the plants in the table would survive winter temperatures of below −20°C? (1)

c) Which plant matches the following description? It has a hardiness score of 4, requires sunlight and flowers in summer. (1)

d) Identify TWO features from the table that begonia and gladiolus have in common. (2)

e) Give THREE features from the table by which hyacinth and snowdrop differ. (3)

f) Imagine you have a garden that gets plenty of sun but its soil temperature can drop to −15°C in winter. Which plants in the table could be successfully planted to give a display of flowers in the spring? (2)

g) Copy the bar graph shown in figure 10.42 and complete it using data from the table. (2)

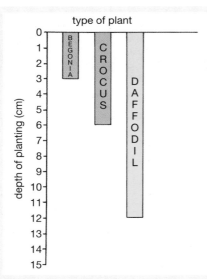

**Figure 10.42**

**5** Some gardeners plant each potato tuber in a separate hole. Figure 10.43 shows some early (E) and some main crop (M) potatoes correctly planted.

a) With reference to figure 10.43, decide whether each of the following statements is true or false. (6)

   (i) The top of an early potato tuber should be 5 cm below the soil surface.

   (ii) The bottom of an early potato tuber must be 15 cm below the soil surface.

   (iii) The top of a main crop tuber should be 10 cm below the soil surface.

   (iv) The bottom of a main crop tuber must be 5 cm below the soil surface.

   (v) Early tubers should be planted 30 cm apart.

   (vi) Main crop tubers should be planted 40 cm apart.

b) A gardener was given a batch of main crop potatoes of varying length but all twice as thick as normal. An expert advised him to plant them at the usual depth but 45 cm apart. Draw a diagram to show the arrangement that he should use for three of these potato tubers. Make your diagram to the same scale as figure 10.43. (3)

c) Another gardener began with 12 early potato tubers which he planted correctly. He got eight tubers from each plant. He set about building up his stock by planting all these tubers. At the same rate of production, how many new potato tubers should he have the following year at harvest time? (1)

**6** Read the following set of instructions for taking cuttings from chrysanthemum plants.

In spring, chrysanthemum plants produce a cluster of new shoots at the base of an old plant. These new shoots can be used as

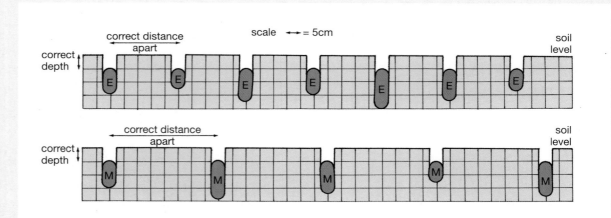

**Figure 10.43**

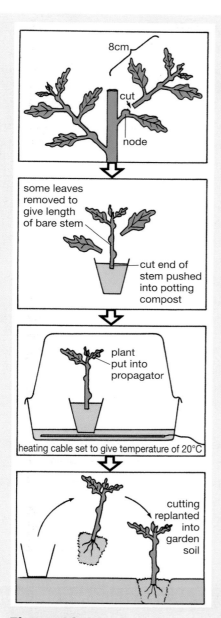

**Figure 10.44**

cuttings. Each shoot should be cut just below a node giving a total shoot length of 4–5 cm. Older leaves should be removed from the lower half of the cutting. The cut end of the stem should be dipped in rooting powder and planted in a hole made using a dibber in cutting compost.

As much of the stem should be below the ground as above it. The cutting in its pot should then be put into a propagator with a bottom heat giving a temperature of 10°C to

encourage root growth. After several weeks, each plant should be ready to be planted out of doors in garden soil.

Figure 10.44 shows a pupil's attempt at following this procedure.

Give FIVE possible reasons why her cutting did not root properly. (5)

7 A gardener wanted to find out the best conditions for propagating mother-in-law's tongue plant from leaf cuttings. Figure 10.45 shows pieces of leaf being prepared and planted. Eight pots (A–H) were prepared in this way and left set up as shown in figure 10.46.

Which TWO parts of the experiment shown on page 262 should be compared to investigate the effect of:
a) age of leaf at 20°C in sunlight? (1)
b) age of leaf at 10°C in shade? (1)
c) sun or shade on old leaves at 10°C? (1)
d) sun or shade on young leaves at 20°C? (1)
e) temperature on young leaves in shade? (1)
f) temperature on old leaves in sun? (1)

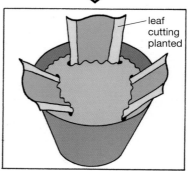

**Figure 10.45**

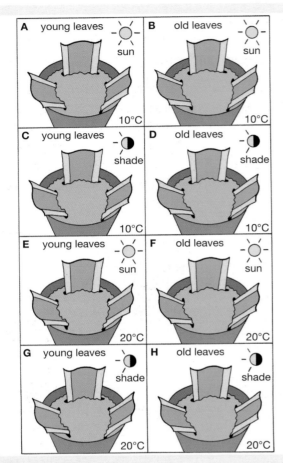

**Figure 10.46**

8 The graph in figure 10.47 shows the results from an experiment where six groups of ten cuttings from a type of plant were each placed in a different concentration of rooting powder solution.

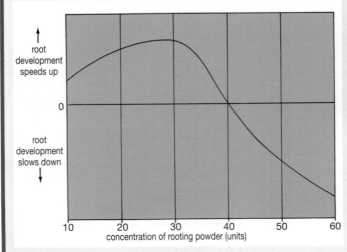

**Figure 10.47**

a) What was the one variable factor being investigated in this experiment? (1)

b) Give TWO other variables that must be kept constant in this experiment to make it valid. (2)

c) Which concentration of rooting powder neither slowed down nor speeded up root development? (1)

d) Which concentration of rooting powder was best at speeding up the development of roots? (1)

e) Describe the control experiment that should have been run. (2)

f) What could be done to make the results of the experiment more reliable? (1)

9 A large selection of different types of flowering plant were investigated to find out which methods could be used to propagate them. The results are shown in table 10.2.

a) What was the total number of plant types investigated? (1)

b) Copy and complete the table by converting the data in the central column into percentages and entering them in the right hand column. (2)

c) Draw a pie chart of these percentages. (3)

d) Shade in the parts of the pie that represent natural methods of vegetative propagation

| method of propagation | number of plant types pro-pagated by this means | percentage number of plant types pro-pagated by this means |
| --- | --- | --- |
| bulb formation | 18 | |
| tuber formation | 0 | |
| plantlet formation | 30 | |
| taking cuttings | 60 | |
| layering | 12 | |

**Table 10.2**

and use a different colour for those that represent artificial methods. (1)

**10** Read the following passage and answer the questions that follow it.

### JUICY FRUIT GALORE

A small garden can produce a large variety of juicy fruit for the table. The plants that produce the fruit can often be propagated. For strawberries, runners are the natural method of increasing the supply. If they are pegged down in summer, this makes them root quickly and produce a lot of fruit the next year.

Raspberry bushes propagate themselves by producing many side shoots called canes alongside the parent plant. The young canes can be dug up and transplanted in autumn. Raspberries can also be propagated artificially by taking root cuttings in autumn.

The part of the rhubarb plant that people eat is the leaf stalk. Rhubarb can be propagated by dividing the roots in spring or autumn. Blueberry plants can be propagated artificially in mid summer by taking softwood stem cuttings. Blackcurrant bushes, on the other hand, can be increased in number by taking hardwood stem cuttings in autumn.

a) Copy and complete table 10.3 using the information in the passage. (10)

b) The last sentence in the passage includes the words *increased in number*. Give ONE word, used elsewhere in the passage, that also means this. (1)

c) A true fruit is formed from a flower. Which of the examples named in the table is *not* a true fruit? Explain your answer. (2)

**11** *Begonia rex* has large multicoloured leaves. The plant can be propagated by taking leaf cuttings. Two ways of doing this are shown in figure 10.48. A grower took six *Begonia rex* leaves and, following method 1, cut each in five places before planting them. He

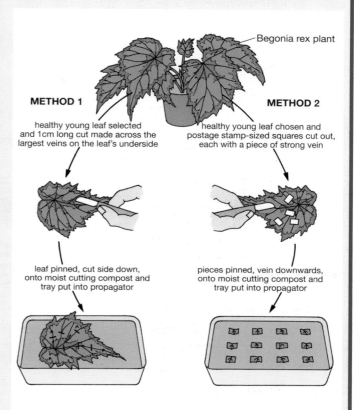

**Figure 10.48**

| fruit | method of propagation | natural or artificial | when to carry out propagation |
|---|---|---|---|
| blueberry | softwood stem cuttings | artificial | summer |
| | root cuttings | | autumn |
| blackcurrant | | | |
| | | artificial | |
| | runners | | summer |
| rhubarb | | | |

**Table 10.3**

also took five other leaves from *Begonia rex* and, following method 2, cut out five squares from each which he planted.

a) If each cut on a *Begonia rex* leaf normally produces one new plant, how many plants could the grower expect from method 1? (1)

b) If each square of leaf produces one new plant, how many plants could he expect from method 2? (1)

c) His actual results were 21 new plants from method 1 and 19 new plants from method 2.

(i) What was his percentage success rate for method 1?

(ii) What was his percentage success rate for method 2?

(iii) Which method produced the higher percentage number of new plants? (3)

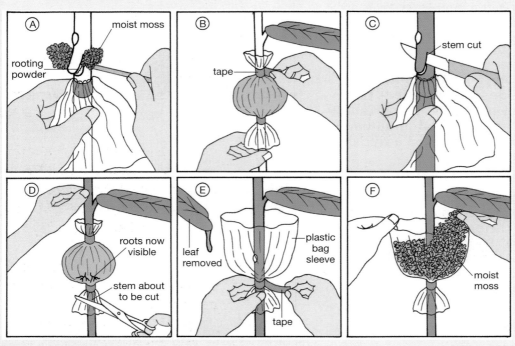

**Figure 10.49**

## AIR LAYERING

Air layering is a method of propagation that gardeners have used for thousands of years on plants that are very difficult or impossible to grow from cuttings. Whereas simple layering involves bringing the plant's stem down and into contact with moist soil, air layering involves bringing the moist growing medium up to meet the plant.

This is done as follows. A healthy young shoot is chosen. Four or five of its leaves are removed and a plastic sleeve is fixed at the lower end of the cleared part. The stem is then prepared by making an upward angled cut into it about 5 cm long. The wound is dusted with rooting powder. Moist Sphagnum moss is packed between the cut surfaces. More moss is packed around the wound and kept in place by the plastic sleeve taped at top and bottom to prevent moisture escaping. When new roots have appeared, the stem is cut below the root ball and the young plant potted in compost.

**12** Read the passage opposite and answer the questions.

a) Figure 10.49 shows the procedure for carrying out air layering but the steps are in a mixed-up order. Arrange them into the correct sequence. (1)

b) Give the main difference between simple layering and air layering. (2)

c) Why is the plastic sleeve taped at the top and bottom? (1)

d) Why do gardeners carry out air layering instead of taking cuttings? (1)

## What you should know

| | | |
|---|---|---|
| artificial | humidity | plantlet |
| bulbs | layering | plastic |
| cut | leaf | propagation |
| cuttings | leaves | rooting |
| difficult | mist | roots |
| disease | naturally | runners |
| energy | node | stem |
| faster | offsets | tubers |
| frost | parent | vegetative |
| heat | pegged | water |

**Table 10.4** *Word bank for chapter 10*

1  The process by which growers increase their supply of a type of plant is called _____.

2  Propagation which does not involve flowers, seeds or sexual reproduction is called _____ propagation.

3  Some methods of vegetative propagation make use of propagation structures formed _____ by the plant.

4  Natural methods include using food storage organs such as _____ and _____ and making use of plantlets. A _____ is a miniature plant attached to a _____ plant. Plantlets are found at the ends of _____ (horizontal stems) and at the edges of some plants' leaves. Others take the form of side shoots called _____ formed at the base of the parent plant.

5  Some methods of vegetative propagation can be used to increase the supply of the plant by _____ means. Plants are often propagated by taking _____. When a _____ cutting is taken, the stem is normally cut just below the _____ because the node is a point of growth. It will respond to the cut (wound) by developing _____. Some plants can be propagated using _____ cuttings.

6  The development of roots by a cutting is often promoted by applying a chemical called _____ powder to the cut surface.

7  A stem cutting with many leaves may lose too much water and die before its roots develop. To prevent this happening, most of its _____ are removed or it is given conditions with an increased level of _____. This can be done by enclosing it in a _____ bag or by putting it in a _____ propagator.

8  Plants can also be propagated artificially by _____. This means pegging the stem down in the soil until roots form at the nodes. Sometimes the stem is _____, part the way through, below a node before being _____ down.

9  An advantage of layering is that it can be used with plants that are _____ or impossible to propagate from cuttings.

10  Supplying the plants being propagated with _____ may be of advantage if it leads to _____ growth and/or prevents _____ damage. However these benefits have to be balanced against the possible disadvantages. These include wilting of plants (caused by high _____ loss in warm conditions), rapid spread of _____ and high _____ costs.

## 11  Plant production

## Conditions for plant growth

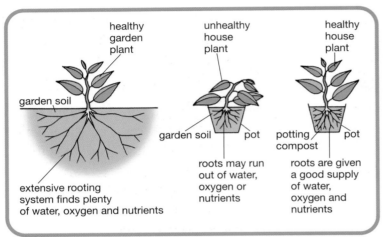

**Figure 11.1**  *Natural and artificial growing conditions*

### Growing medium

For healthy growth a plant needs to take in **water** and **nutrients** through its roots. Its roots also need to get **oxygen** from air spaces in the soil for respiration. In its natural environment, a plant growing in the ground has plenty of room to spread out its roots and take in water, oxygen and nutrients from a large volume of soil (see figure 11.1).

Under *artificial* conditions, the same plant grown in a pot as a house plant only has a small volume of soil for its roots to use. If this were ordinary, untreated garden soil, it would probably not hold enough water or air for the plant's needs. In addition, it would soon run out of nutrients and the plant would not grow properly.

### Compost

Pot plants need a growing medium that holds plenty of water, lets air in for the roots to respire and is rich in nutrients. It is for these reasons that house plants are normally grown in a specially made up growing medium called **potting compost**. (*Note*: This is different from garden compost in a compost heap which is dead or decaying leaf and food material used to improve garden soil.)

**Figure 11.2**  *Materials for making compost*

**Figure 11.3**  *Peat and peat substitutes*

| material | property | notes |
|---|---|---|
| loam (good garden soil) | provides the compost with nutrients and a basic structure | it must be sterilised before use by heat treatment to kill pests, weed seeds and harmful micro-organisms |
| peat | improves the compost's ability to hold (retain) water | it is dead bog moss which is a non-renewable resource so it may be replaced by a peat substitute (see figure 11.3) |
| grit and sand | increase the compost's air content and let excess water drain away easily | fairly large particles are needed; very fine sand is unsuitable because it tends to block air spaces |
| perlite | increases the compost's air content and lets excess water drain away easily | made of small white granules of volcanic minerals |
| peat substitutes | improve the compost's ability to hold (retain) water | these are normally cocoa shells, bark and coir (hairy fibre from coconut) |
| fertiliser | increases the compost's concentration of nutrients | this contains nutrients such as nitrogen (N), phosphorus (P) and potassium (K) that are needed for healthy plant growth |

**Table 11.1** *Properties of materials used in potting composts*

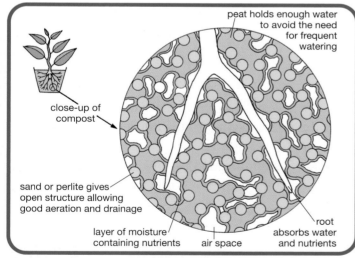

**Figure 11.4** *Loamless potting compost*

## Loam and loamless composts

Table 11.1 and figures 11.2 and 11.3 show some of the materials used to make up potting composts and their properties. Each compost contains a selection of these materials. If it contains loam (good garden soil) as one of its ingredients, it is called a **loam-based** compost. If it does not contain loam, it is called a **loamless** compost. Table 11.2 compares these two types of compost.

## Potting and rooting composts

A simplified version of loamless **potting compost** in use is shown in figure 11.4. Such compost is

| | loam-based compost | loamless compost |
|---|---|---|
| main material | sterilised loam | peat (or substitute) |
| other materials (in smaller quantities) | sand (or perlite) fertiliser peat (or substitute) | sand (or perlite) fertiliser |

**Table 11.2** *Comparison of loam-based and loamless compost*

**Figure 11.5** *Different types of compost*

ideal for a pot plant with an established rooting system. However, if a gardener wants to encourage cuttings to develop roots, s/he would probably use a special type of compost called **rooting compost**. This has a higher content of sand (or perlite) in it to give very good drainage. Therefore the buried end of the cutting does not rot in over-wet compost.

Many different types of compost can be made by varying the ingredients and their proportions. Figure 11.5 shows some examples of commercially produced composts. Some of these are loam-based (e.g. John Innes No. 3) and others are loamless (e.g. African violet compost).

## Activity

### Making rooting and potting compost

You need
- peat
- sand (or perlite)
- multipurpose fertiliser (powder or pellets)
- trowel
- bucket
- watering can and water
- beaker (1 litre, plastic)
- spatula
- access to electronic balance

What to do
Rooting compost
1 Using the plastic beaker, measure out 1 litre of peat and 1 litre of sand into the bucket.
2 Mix the two using a trowel.
3 Add water slowly and carefully from the watering can to moisten the compost but not soak it.
4 Mix the ingredients thoroughly and leave for 24 hours before using to root cuttings.

Potting compost
1 Using the plastic beaker, measure out 3 litres of peat and 1 litre of sand into the bucket.
2 Read the instructions on the packet of fertiliser and weigh out an appropriate quantity.
3 Add fertiliser to the bucket and mix the materials using the trowel.
4 Add water slowly and carefully from the watering can to moisten the compost but not soak it.
5 Mix the ingredients thoroughly and leave for 24 hours before using to repot plants.

# Investigating the water-holding capacity and drainage of two types of compost

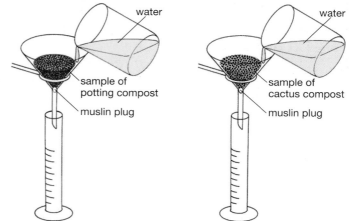

**Figure 11.6** *Investigating water-holding capacity and drainage of composts*

## You need
- multipurpose potting compost
- cactus compost
- two filter funnels
- muslin
- two measuring cylinders (100 cm³)
- two beakers (100 cm³) containing 50 cm³ of water
- two clamp stands with metal ring
- stopclock

## What to do
1 Place a plug of muslin in each filter funnel.
2 Set up each filter funnel in a ring held by a clamp as shown in figure 11.6.
3 Half-fill one filter funnel with multipurpose compost and the other with cactus compost.
4 Place a measuring cylinder under each filter funnel.
5 Slowly add 50 cm³ of water to each compost sample.
6 Start the clock and allow five minutes before reading the volume of water in each measuring cylinder.
7 Write a brief report by doing the following:
   a) Make up a title for this investigation and write it into your book.
   b) Draw a simple labelled diagram of the apparatus.
   c) Copy and complete table 11.3.
   d) Present your results as a bar chart.
   e) Answer the following questions:
      (i)   What was the one variable factor in this experiment?
      (ii)  Name three factors that were kept the same to allow a fair and valid comparison of the results to be made.
      (iii) Which type of compost allows water to drain through it more easily?
      (iv)  Which type of compost has the higher water-holding capacity?
   f) Examine fresh samples of the two composts and explain your answers.

| compost type | volume of water added (cm³) | volume of water in measuring cylinder after five minutes of drainage (cm³) | volume of water held by compost (cm³) | percentage water-holding capacity of compost |
|---|---|---|---|---|
| multipurpose | 50 | | | |
| cactus | 50 | | | |

**Table 11.3**  *Results*

## Nutrients

A plant needs certain **nutrients** for healthy growth. These are absorbed by its roots from the surrounding growing medium. Nutrients (sometimes called **minerals**) are chemicals such as **nitrogen** (N), **phosphorus** (P) and **potassium** (K). Since each of these plays an important role in plant growth (see table 11.4) and is required in fairly large quantities, it is called a **major plant nutrient**. Other chemicals are important too. Some of them are only needed in tiny amounts and are called **trace nutrients** (elements). Two examples are **iron** and **magnesium**.

| nutrient | | importance of nutrient to plant |
|---|---|---|
| **chemical name** | **chemical symbol** | |
| nitrogen | N | promotes leaf growth |
| phosphorus | P | promotes root growth |
| potassium | K | promotes growth of flowers and fruit |

**Table 11.4**  *Role of major plant nutrients*

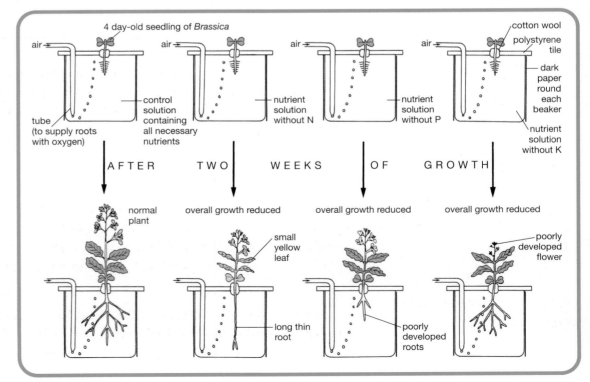

**Figure 11.7**  *Water culture experiment*

## Water culture experiment

The importance of each of the three major plant nutrients can be investigated by setting up a **water culture** experiment (see figure 11.7). The first beaker is the **control** experiment. It contains all the nutrients needed for plant growth. Each of the other three beakers contains a solution from which one of the three nutrients has been left out.

The glass beakers used in this water culture experiment are first rinsed with concentrated acid to remove any traces of chemicals on their inner surface. Each is surrounded by dark paper to stop green algae growing. Although only one plant (fast-growing *Brassica*) is shown in the diagram for each beaker, several plants are used each time. This makes the results **more reliable** than they would be if only one plant was used.

After about 14 days, each plant lacking an essential nutrient is found to show certain signs of poor health (**deficiency symptoms**) when compared with the control. The plants lacking nutrients all show **overall reduced growth**. In addition, the ones with **no nitrogen** have **poor leaves**, the ones with **no phosphorus** have **poor roots** and the ones with **no potassium** have **poor flowers**.

## Fertilisers

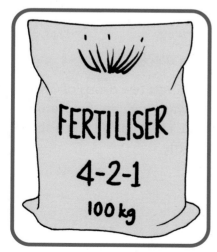

**Figure 11.8** *Mineral ratio of fertiliser*

A **fertiliser** is a substance containing extra supplies of the nutrients needed by growing plants. If the plant's growing medium is short of nutrients, fertiliser can be added to enrich it with the necessary chemicals that the plant needs.

### Mineral ratios

The quantity of N, P and K present in a commercially produced fertiliser varies from product to product. The proportions of these chemicals contained in the fertiliser are shown on the pack as the **mineral ratio**. This always refers to the chemicals in the order N, P and K.

A mineral ratio of 4% N : 2% P : 1% K would be written 4:2:1 or 4-2-1 for short. The 100 kg bag of fertiliser shown in figure 11.8 would contain 4% N (4 kg), 2% P (2 kg) and 1% K (1 kg).

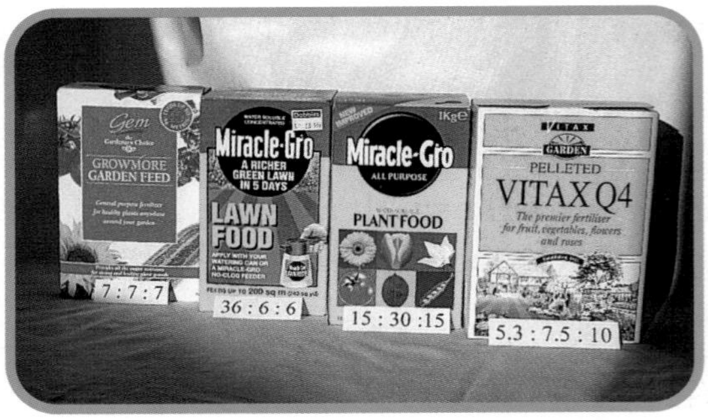


\Figure 11.9 *Various mineral ratios*

## Types of fertiliser

The mineral ratio varies from product to product (see figure 11.9). Some fertilisers contain a balanced mixture of nutrients, e.g. 7-7-7, and are called **multipurpose fertilisers**.

Some fertilisers contain more of one nutrient and less of the other two because they are designed to do a particular job. For example, a fertiliser containing 36-6-6 has a high proportion of **N** to boost **leaf** growth. A fertiliser with a nutrient ratio of 15-30-15 has a high proportion of **P** to promote **root** growth. A fertiliser labelled 5.3-7.5-10 is rich in **K** for **flower** and **fruit** production. In each of these examples, the fertiliser also contains the other nutrients, though in smaller quantities, to ensure good growth of the other plant parts in addition to the one being targeted.

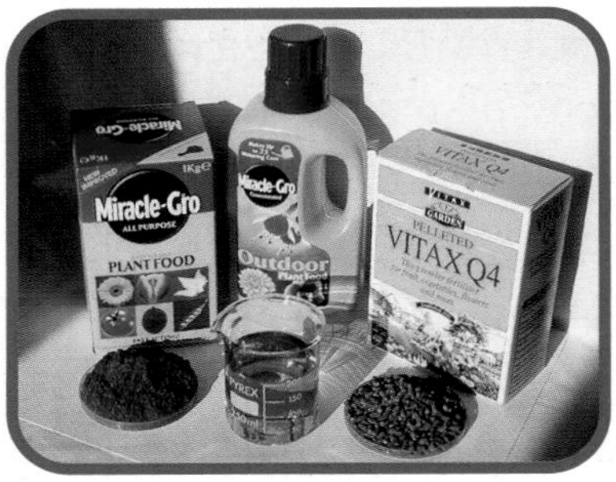


**Figure 11.10** *Different forms of fertiliser*

## Applying fertiliser

Fertilisers are produced in the form of powder, liquid and granules as shown in figure 11.10.

### Powder

Fertiliser powder is often applied by simply scattering it over the surface of the soil or sprinkling it around individual plants. However, sometimes the powder must be dissolved in water to give a particular concentration before being applied.

### Liquid

Some fertilisers are produced in liquid form ready to be applied directly to the plant's growing medium.

#### Liquid feeding

People often 'feed' their house plants by adding a few drops of concentrated fertiliser to the water in their watering can. An example of this type of chemical is shown in figure 11.11a. Figure 11.11b shows a close-up of the label on the back of the bottle.

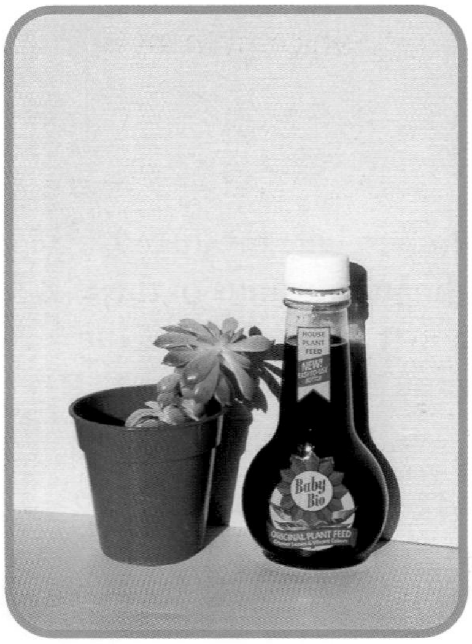


**Figure 11.11a** *Liquid fertiliser*

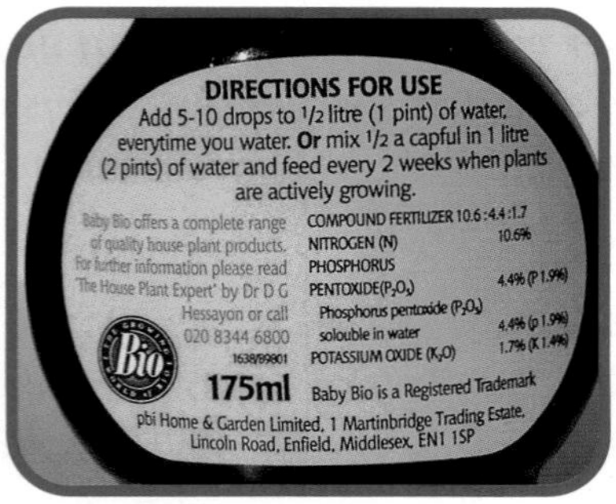


**Figure 11.11b** *Close-up of label*

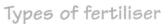

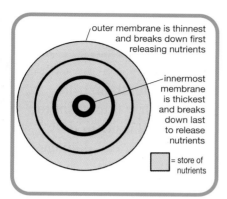

outer membrane is thinnest and breaks down first releasing nutrients

innermost membrane is thickest and breaks down last to release nutrients

☐ = store of nutrients

**Figure 11.12** *Slow-release fertiliser granule*

Dissolved fertiliser and liquid feed give a quick-acting boost to plants and let the person decide when to add the nutrients. However, out of doors in the garden or farm, heavy rain can quickly wash the nutrients out of the soil before the plants have had time to absorb them.

### Granules

Granules are small pellets designed to release nutrients slowly. They often have **membranes** of varying thickness (see figure 11.12) that gradually break down and release the nutrients stored inside them.

The granules do not have all their nutrients washed out by a few heavy showers of rain and the plants continue to be supplied with nutrients for months or even years by the same pellets.

## Testing your knowledge

1 Table 11.5 gives the ingredients needed to make up two types of potting compost.
   a) Match X and Y with *loamless compost* and *loam-based compost.* (1)
   b) What is loam? (1)
   c) Why does the garden soil used for compost X need to be sterilised? (1)
   d) (i) Give ONE reason why peat is used as a basic ingredient of potting compost.
      (ii) Name a peat substitute. (2)
   e) (i) Give ONE reason for using sand in potting compost.
      (ii Name a substance that can be used instead of sand. (2)

2 a) Give the name and chemical symbol of each of the three main nutrients needed by plants. (6)
   b) (i) If a fertiliser has a nutrient ratio 20-1-1, which nutrient is present in the highest proportion?
      (ii) What effect will this fertiliser have on plant growth? (2)
   c) If a fertiliser has the nutrient ratio 6-6-6, what effect will it have on plant growth? (1)

| compost X | compost Y |
|---|---|
| seven parts of sterilised garden soil | three parts of coarse peat |
| three parts of peat | one part of sand |
| one part of sand | fertiliser |
| fertiliser | |

**Table 11.5**

# Watering

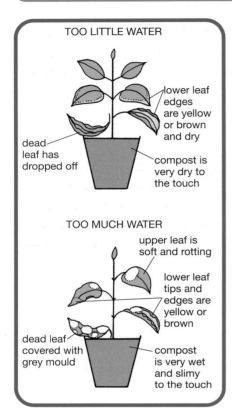

TOO LITTLE WATER

lower leaf edges are yellow or brown and dry

dead leaf has dropped off

compost is very dry to the touch

TOO MUCH WATER

upper leaf is soft and rotting

lower leaf tips and edges are yellow or brown

dead leaf covered with grey mould

compost is very wet and slimy to the touch

**Figure 11.13** *Too little or too much water*

Plants need water to give them support and prevent them from wilting. Different plants need different amounts. It is important to find out if a plant needs plenty of water (e.g. busy Lizzie) or only limited amounts (e.g. barrel cactus). Too much water can damage a plant just as easily as too little (see figure 11.13).

Most house plants should only be watered when the surface of the compost has become dry and crumbly to the touch. After watering, the compost should be moist but not saturated because roots need air as well as water.

## Methods of watering house plants

Different methods of watering suit different plants as shown in figures 11.14 and 11.15.

### Watering from above

For most plants, the water is poured directly from a **watering can** onto the compost, briefly filling most of the space between the surface of the compost and the rim of the pot. The water should quickly drain away. More water is slowly added until a small quantity appears in the drainage saucer. If a large volume of water gathers in the drainage saucer, it should be emptied so that the plant pot is not left standing in water.

Bromeliads are unusual plants that grow in the branches of trees in tropical rain forests. They get most of their water from rain that falls

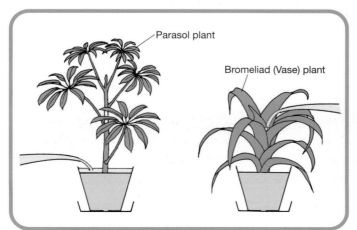

Parasol plant

Bromeliad (Vase) plant

**Figure 11.14** *Watering from above*

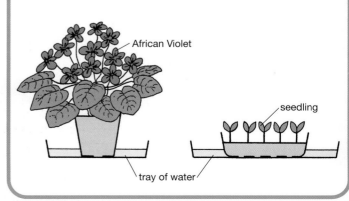

African Violet

seedling

tray of water

**Figure 11.15** *Watering from below*

into the 'vase' formed by their leaves. They should be watered from above by pouring a little water into their 'vases'.

### Watering from below

Some plants (e.g. African violet and young delicate seedlings) must be watered from below. This can be done by standing the pot or tray in shallow water for about 15 minutes until the roots have absorbed as much water as they need.

## Watering garden plants

A large watering can is ideal for watering a small garden. However in a large garden where many of the plants are situated far away from the source of water, a **hose** is more useful. It also lets the gardener easily irrigate areas that need large supplies of water frequently. A sprinkler attached to a hose is useful for watering a large lawn during a dry spell.

### Activity

## Watering (assessment)

### You need
- potted geranium plant in drainage saucer
- potted African violet plant
- small tray of seedlings
- watering can
- access to trough or large tray of water 2 cm deep
- stopclock

### What to do
1 Read *Methods of watering house plants* on page 274.
2 Collect the three containers of plants that are to be watered.
3 Consider the equipment that is available and decide how you are going to use it.
4 Ask your teacher to assess you as you water the plants.

## Automatic methods of watering plants

Plants are normally watered by the grower using a watering can. However there are times when s/he is not available and the plants still need to be watered. Under these circumstances, **automatic methods** of watering are used.

### Automatic irrigation

This is also known as **trickle irrigation**. It supplies the plant with water from above. It consists of a plastic pipe leading to a series of

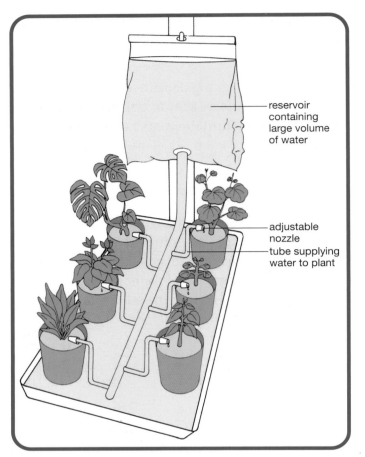

**Figure 11.16** *Automatic irrigation*

reservoir containing large volume of water

adjustable nozzle

tube supplying water to plant

reservoir containing large volume of water

gutter filled with water

capillary matting

**Figure 11.17** *Use of capillary matting*

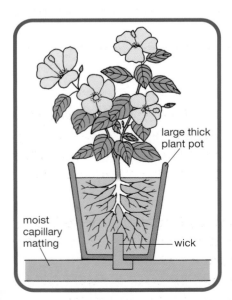

large thick plant pot

moist capillary matting

wick

**Figure 11.18** *Use of wick*

narrower tubes as shown in figure 11.16. Each narrow tube has an **adjustable nozzle** at its end that releases the water.

The tubes are arranged each with a nozzle in a plant pot. The system gets its water from a reservoir (or directly from the mains). A trickle of water at a rate suited to the needs of each type of plant is selected by adjusting the appropriate nozzle.

## Capillary matting

**Capillary matting** is a man-made material that draws up water and stays moist. It is often used in the greenhouse to supply plants with water from below. The bench is first lined with polythene sheeting and then a layer of capillary matting is placed on top. The capillary matting is kept moist by placing its edge in a reservoir of water such as a length of gutter (see figure 11.17).

The plant pots or seed trays are then placed on the capillary matting. This arrangement normally gives enough contact between the matting and the compost in the pots or trays for water to be absorbed. However it may not work for large, thick pots. These need to be supplied with a **wick** (a thin strip of capillary matting) as shown

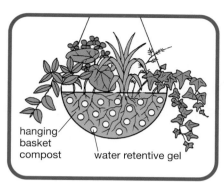

hanging
basket
compost    water retentive gel

**Figure 11.19**  *Hanging basket*

in figure 11.18. The wick is stuck into the pot's drainage hole to bridge the gap and allow water to pass from the capillary matting to the compost.

### Water retentive gel

A **water retentive gel** is a chemical that is able to absorb and hold several hundred times its own weight of water. This type of chemical is added to compost to cut down the need for frequent watering e.g. in a hanging basket (see figure 11.19).

## Activity

### ACTIVITY Designing and making a watering system for house plants

#### You need
- See apparatus in figure 11.20

#### What to do
1 Design a watering system by drawing a simple labelled diagram to show the equipment in use.
2 Show your plan to the teacher for approval. (If you are stuck and need help, see figure 11.60 on page 301 at the end of the chapter.)
3 Set up the watering system and, if possible, keep it set up for two weeks to see if it works.

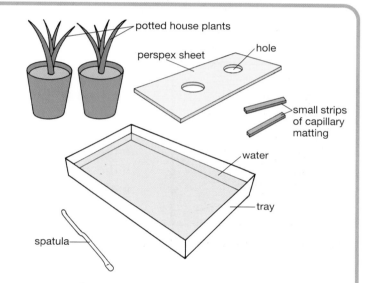

potted house plants
perspex sheet
hole
small strips of capillary matting
water
tray
spatula

**Figure 11.20**  *Apparatus for designing a watering system*

## Environmental conditions

### Temperature

Each type of plant has an **optimum range** of temperature. This means the range within which it grows best. A spider plant, for example, only grows well between 10 and 30°C. If the temperature drops below 5°C or rises above 35°C, the plant does not grow well at all and may die.

### Controlling temperature

When people make their living by growing plants, they have to find ways of controlling the plants' environment. One of the best ways of doing this is to grow the plants (e.g. tomato plants) in a **greenhouse** (see figure 11.21) or in its cheaper alternative, a walk-in **polythene tunnel** made of heavy duty plastic sheeting draped over metal supports.

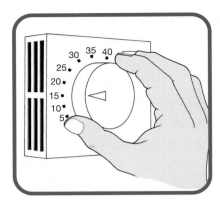

**Figure 11.22**
*Thermostatic control of a greenhouse*

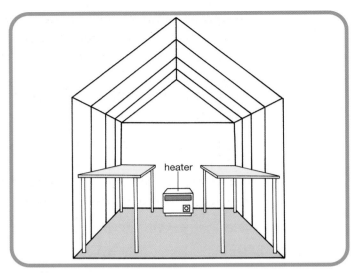

**Figure 11.21**  *A greenhouse*

The temperature in the greenhouse or polythene tunnel is kept at a suitable level using a **heater**. Often this is an electric heater controlled by a **thermostat**. The grower sets the thermostat at a certain temperature (see figure 11.22) which would probably be somewhere in the middle of the plant's optimum range. When the temperature reaches this level, the thermostat switches the heater off.

When the greenhouse begins to cool down, the thermostat switches the heater back on again until the required temperature is reached and so on.

## Maximum and minimum thermometer

If growers wish to check the range of temperature that their plants have been exposed to during the previous 24 hours, they can use a **maximum and minimum thermometer**.

A maximum and minimum thermometer is U-shaped as shown in figure 11.23. One arm of the U records the maximum temperature that occurred during the previous 24 hours, the other records the minimum temperature that occurred during the previous 24 hours.

Look at the thermometer shown in the diagram. During the last 24 hours, the mercury in the arm on the right-hand side moved along the tube until a maximum temperature was reached. As the mercury moved, it also pushed the **metal marker** along. Later, when the temperature dropped, the mercury moved back along the tube leaving the metal marker behind. So the *bottom* of the metal marker shows the *maximum point* reached by the mercury on the scale. In this case 30°C was the maximum temperature.

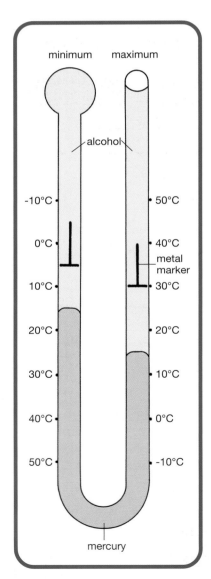

**Figure 11.23**  *Maximum and minimum thermometer*

Now look at the left side of the thermometer in the diagram. During the last 24 hours, the mercury moved along the scale until a minimum temperature was reached. Again it also pushed a metal marker along. Later, when the temperature rose, the mercury moved back along the tube but the metal marker stayed behind. So the *bottom* of this marker shows the *minimum point* reached by the mercury on the scale. In this case 5°C was the minimum temperature reached.

Both arms of the thermometer also show the **present temperature**, in this case 15°C. A **magnet** is used to move the metal markers back into contact with the mercury at an agreed time each day ready to begin the next set of readings for a 24-hour period.

## Activity

### Using a maximum and minimum thermometer

*You need*

- access to a maximum and minimum thermometer that has been reset and then left in a greenhouse or science laboratory for 24 hours

| day number | maximum temperature (°C) | minimum temperature (°C) |
|---|---|---|
| 1 | | |
| 2 | | |
| 3 | | |
| 4 | | |

**Table 11.6**   *Results for maximum and minimum thermometer*

*What to do*

**1** Read *Maximum and minimum thermometer* on page 278.
**2** Put the above heading in your book.
**3** Copy table 11.6.
**4** Read and record the maximum and minimum temperatures on the thermometer for the previous 24 hours and enter them in the table for day 1.
**5** Once everyone has taken readings, choose one person to use the magnet to move the two markers back into contact with the mercury on both sides of the thermometer.
**6** Leave the thermometer in the same place as before for 24 hours.
**7** Repeat the procedure for days 2, 3 and 4 and complete the table as you go along.
**8** Copy the two axes shown in figure 11.24 onto a sheet of graph paper and plot two line graphs of the results on the same sheet of graph paper. (Invent a colour code.)

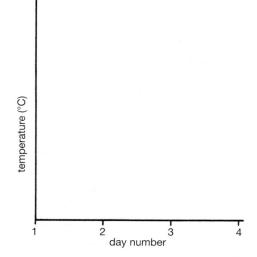

**Figure 11.24**   *Axes for graph*

## Humidity

**Humidity** is a measure of the quantity of **water vapour** in the air. Plants lose water by evaporation from their leaf surfaces. This loss of water has a cooling effect similar to that of sweating in humans. If the air is very humid (damp), hardly any water will evaporate from the plant. If the air is very dry, the plant may lose too much water and wilt or even die. Different levels of humidity suit different types of plant.

### Relative humidity

When scientists want to compare the humidity of different air samples, they measure **relative humidity**.

relative humidity of an air sample =

$$\frac{\text{quantity of water vapour in that air}}{\text{quantity of water vapour in saturated air}} \times 100$$

For example, a relative humidity of 75% would be typical of a *humid* atmosphere whereas a relative humidity of 35% would indicate a *dry* atmosphere.

The relative humidity of the air around us varies from day to day and even from hour to hour. The higher the temperature, the more water vapour the air is able to hold.

### Monitoring relative humidity

A **wet and dry bulb hygrometer** is made up of two ordinary thermometers mounted on a stand as shown in figure 11.25. Since one thermometer has its bulb wrapped in wet muslin while the other has its bulb left untreated, the equipment is sometimes called '*wet and dry bulb thermometers*'. The wet bulb will show a lower temperature than the dry bulb due to evaporation from the wet bulb.

Relative humidity can be worked out by doing the following:

- read the dry bulb's temperature
- read the wet bulb's temperature
- subtract the wet reading from the dry reading to give the **depression of wet bulb reading** as in the equation:

*depression of wet bulb reading = dry reading – wet reading* (For example, this would be 3°C for the apparatus shown in figure 11.25.)

- refer to a table of relative humidities (see table 11.7) and read off the value. (For example a dry bulb temperature of 18°C and a depression of wet bulb reading of 3°C means a relative humidity of 71%.)

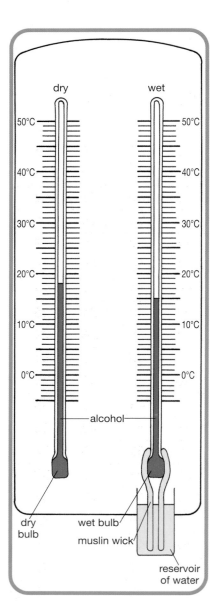

**Figure 11.25** *A wet and dry bulb hygrometer*

| dry bulb temperature (°C) | relative humidity (%) depression of wet bulb (°C) | | | | | | | |
|---|---|---|---|---|---|---|---|---|
| | **1** | **2** | **3** | **4** | **5** | **6** | **7** | **8** |
| 1 | 81 | | | | | | | |
| 2 | 82 | 64 | | | | | | |
| 3 | 83 | 66 | 49 | | | | | |
| 4 | 83 | 67 | 51 | 36 | | | | |
| 5 | 84 | 69 | 53 | 39 | 24 | | | |
| 6 | 85 | 70 | 55 | 41 | 28 | 14 | | |
| 7 | 85 | 71 | 57 | 44 | 31 | 18 | 5 | |
| 8 | 86 | 72 | 59 | 46 | 33 | 21 | 9 | |
| 9 | 86 | 73 | 61 | 48 | 36 | 24 | 13 | 2 |
| 10 | 87 | 74 | 62 | 50 | 38 | 27 | 16 | 5 |
| 11 | 87 | 75 | 63 | 52 | 41 | 30 | 19 | 9 |
| 12 | 88 | 76 | 65 | 54 | 43 | 32 | 22 | 12 |
| 13 | 88 | 77 | 66 | 55 | 45 | 35 | 25 | 16 |
| 14 | 89 | 78 | 67 | 57 | 47 | 37 | 28 | 18 |
| 15 | 89 | 78 | 68 | 58 | 49 | 39 | 30 | 21 |
| 16 | 89 | 79 | 69 | 60 | 50 | 41 | 32 | 24 |
| 17 | 90 | 80 | 70 | 61 | 52 | 43 | 34 | 26 |
| 18 | 90 | 80 | 71 | 62 | 53 | 45 | 36 | 29 |
| 19 | 90 | 81 | 72 | 63 | 55 | 46 | 38 | 31 |
| 20 | 91 | 81 | 73 | 64 | 56 | 48 | 40 | 33 |
| 21 | 91 | 82 | 73 | 65 | 57 | 49 | 42 | 35 |
| 22 | 91 | 82 | 74 | 66 | 58 | 51 | 43 | 36 |
| 23 | 91 | 83 | 75 | 67 | 59 | 52 | 45 | 38 |
| 24 | 91 | 83 | 75 | 68 | 60 | 53 | 46 | 39 |
| 25 | 92 | 84 | 76 | 68 | 61 | 54 | 47 | 41 |
| 26 | 92 | 84 | 76 | 69 | 62 | 55 | 49 | 42 |
| 27 | 92 | 84 | 77 | 70 | 63 | 56 | 50 | 44 |
| 28 | 92 | 85 | 77 | 70 | 64 | 57 | 51 | 45 |
| 29 | 92 | 85 | 78 | 71 | 65 | 58 | 52 | 46 |
| 30 | 93 | 85 | 78 | 72 | 65 | 59 | 53 | 47 |

**Table 11.7** *Table of relative humidities for use with a wet and dry bulb hygrometer*

## Ventilation

When a grower uses a greenhouse, the greenhouse must be ventilated. This is necessary to give the plants fresh air and avoid the build-up of damp stale air which would create ideal conditions for the spread of plant diseases such as grey mould (see page 289).

There are many ways of ventilating a greenhouse. The best methods work automatically. Two of these are described below.

### Automatic vent openers (autovents)

These are attached to the windows and are designed to open and close automatically. Each **autovent** (opener) contains a cylinder of wax (see figure 11.26). The wax expands when the temperature in the greenhouse becomes too high. The expanded wax pushes the cylinder and opens the window.

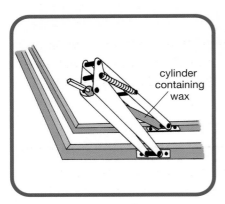

**Figure 11.26** *An autovent*

As the greenhouse becomes ventilated and cools down, the wax contracts and the window closes again.

### Extractor fans

Thermostatically controlled **extractor fans** are often fitted to a greenhouse. When the temperature rises to above the required level, the thermostat switches the fan on and this ventilates the greenhouse. When the air becomes cooler, the thermostat switches the fan off.

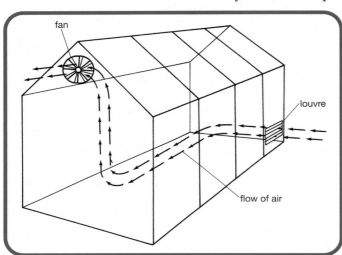

**Figure 11.27** *Effect of a fan*

One popular arrangement is to have a thermostatically controlled fan high up at one end of the greenhouse and a louvre window low down at the other end as shown in figure 11.27. This allows fresh air to flow through the greenhouse when the fan is on.

### Thermostatic control

A greenhouse may have two thermostats. The temperature is kept at the required level by one thermostat switching on the heater to warm up the air in the greenhouse (e.g. on cold days) and the other thermostat switching on the fan to cool down the air in the greenhouse (e.g. on hot days).

## Wind

**Wind** increases the rate at which a plant loses water by evaporation from its leaves. This can have a chilling effect on plants especially young seedlings. These may be permanently damaged by high wind. Sensitive plants can be protected by windbreaks. These can take the form of hedges or man-made structures (see *Protected cultivation*, page 291).

### Wind speed

**Wind speed** can be measured using a hand-held **wind meter** as shown in figure 11.28. The instrument is held facing the wind. The greater the wind speed, the higher the small ball in the central tube rises.

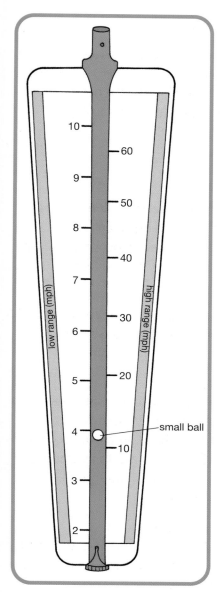

**Figure 11.28** *A wind meter*

(see *Protected cultivation*, page 291).

## Testing your knowledge

**1** a) A plant was found to have upper leaves that were soft and rotting and dead lower leaves covered with grey mould. Had the plant been over-watered or under-watered? (1)

   b) (i) When you are watering a plant from above, how can you tell when you have added enough water?

      (ii) What should you do if a large volume of water gathers in the drainage saucer? (2)

   c) An African violet plant must *not* be watered from above. Describe the method that should be used instead. (2)

**2** a) Using all of the following words and phrases, describe how a system of automatic irrigation works. (5)
adjustable nozzles, plastic tubes, potted plants, reservoir of water, trickle of water

   b) Why is water retentive gel useful in hanging baskets? (1)

**3** Describe how a thermostatically controlled heater regulates the temperature of a greenhouse. (2)

**4** a) What is meant by the term *humidity* of air? (1)

   b) What apparatus can be used to monitor relative humidity of air? (1)

**5** a) Name TWO methods of ventilating a greenhouse. (2)

   b) Why do the plants in a greenhouse need good ventilation? (1)

## Plant maintenance

**Figure 11.29**  *Desert cacti*

For the sake of simplicity, only house plants are going to be considered here. Several different groups of house plant and ways of meeting their individual needs are described below.

### Cacti

A **cactus** is a plant with a fleshy stem that contains **succulent** (water-storing) tissue. The stem of a mature plant may bear **sharp spines**, side shoots and sometimes flowers. However it does not have normal leaves.

#### Desert cacti

Desert cacti are native to semi-desert environments. Figure 11.29 shows two desert cacti being grown as house plants. They need direct sunlight and well-drained soil. They do well when given warm conditions during the summer and cool conditions in winter.

#### Jungle cacti

A few cacti live attached to the trees in jungles. Figure 11.30 shows two examples being grown as house plants. These cacti need a fairly well-lit position that is shaded from direct sunlight, well-drained soil and conditions that are warm during the growing season and cool (but not cold) during the winter.

### Other succulents

A **succulent** plant is one that has thick fleshy parts containing a store of water for survival in a dry environment. In addition to cacti, many

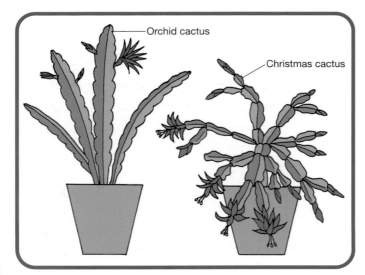

**Figure 11.30**  *Jungle cacti*

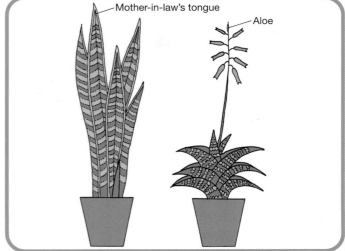

**Figure 11.31**  *Other succulents*

other varieties of plant have fleshy parts. These are *not* cacti because they have normal leaves (not spines).

When grown as house plants, succulent plants such as Aloe (see figure 11.31) that are native to semi-dry environments need a well-lit place with shade from intense sunlight and a well-drained soil. They need warm growing conditions in summer and cool (not cold) conditions in winter.

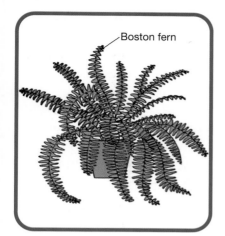

**Figure 11.32**   *A fern*

### Ferns

**Ferns** are non-flowering plants native to damp woodland environments that are not brightly lit. Their leaves grow out of underground stems. When grown as house plants, they thrive in humid (moist) conditions with good indirect light. They like cool (but not cold) conditions. They do not survive in hot, dry environments. An example is shown in figure 11.32.

### Foliage plants

House plants that are grown for the shape and/or colour of their leaves rather than for their flowers are called **foliage plants**. A selection are shown in figure 11.33. Their decorative effect results from a mix of white and other colours amongst the basic green combined with a variety of shapes and sizes.

**Figure 11.33**   *Foliage plants*

Most foliage plants thrive in places that are well lit but out of direct sunlight. They like well-watered (but not over-watered) compost and occasional 'feeds' of fertiliser. Under these conditions, most foliage plants stay healthy and attractive all year (although their growth slows down during the winter).

### Flowering plants

These are plants that are grown for their attractive, decorative **flowers**. Once the flowers die, the plant can often be encouraged to produce more flowers by removing the dead flower heads (see *Dead-heading*, page 288).

**Figure 11.34**   *Fuchsia – a flowering plant*

Eventually no more flowers can be produced but the rest of the plant is still alive and healthy. If it is a perennial plant (one that keeps on growing for several years), it will flower again the next year. In the meantime, it must be kept in the correct growing conditions. In general, flowering house plants need more light than foliage plants. They like well-watered (but not over-watered) compost and an occasional 'feed' of fertiliser.

**Figure 11.35** *A winter flowering plant*

African violet, geranium and fuchsia (see figure 11.34) are examples of plants that often become permanent residents in many people's homes and flower year after year.

## Winter gifts

Some flowering plants such as chrysanthemum and azalea (see figure 11.35) are popular gift plants in the winter. When their flowers die, the remaining healthy plants are often thrown away needlessly. If an azalea is kept moist, cool and brightly lit, it will flower again the following year.

## Methods of maintaining plants

### Pricking out

When seeds have been sown in a tray, the emerging seedlings are often too close together. Before they start competing with one another for root space, water and light, they need to be transplanted. This is done by **pricking them out**. This means transferring the young seedlings from where they were germinated to a new location where they will have more room to grow. Seedlings are ready for pricking out when their first set of leaves open out.

### Activity

## Pricking out seedlings (assessment)

*You need*
- ruler
- small pot (e.g. 80 mm diameter)
- multipurpose compost
- beaker (100 cm³ plastic)
- access to a tray of seedlings
- large spatula
- access to a trough of water
- pencil or dibber

*What to do* (also see figure 11.36)
1 Fill the small plant pot with multipurpose compost.
2 Use the small plastic beaker to level off and gently firm the compost in the pot.

**3** Use a pencil or dibber to make holes in the compost about 20 mm deep and 15 mm wide.

**4** Gently dig up a seedling from the seed tray using the narrow end of the spatula. Carefully lift the seedling by its leaves (to prevent damage to its stem) while at the same time keeping its roots and as much compost as possible in the spatula.

**5** Gently lower the seedling's roots and compost into one of the holes in the plant pot.

**6** Firm the compost around the seedling's stem.

**7** Repeat steps 1–6 for two more seedlings.

**8** Water the pot by sitting it in a trough of water (2 cm deep) for 15 minutes.

**9** Show the pot to your teacher for assessment.

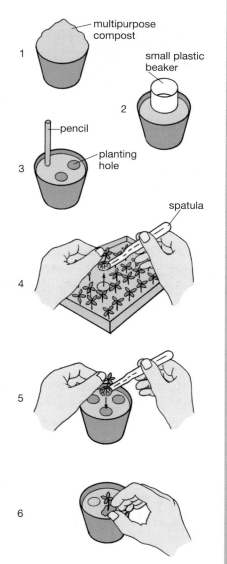

**Figure 11.36** *Pricking out seedlings*

## Potting on

**Potting on** (repotting) is the transfer of a plant from one pot to another larger one. This becomes necessary when the plant has become **pot-bound** (root-bound). Such a plant can be recognised by the following features:

- its growth rate is unusually slow during the growing season;
- its compost dries out very quickly;
- its roots are growing out of the pot's drainage holes.

## Potting on plant (assessment)

### You need

- pot-bound plant in its pot
- selection of pots and drainage saucers
- crocks and potting compost
- watering can and water

### What to do

1 Collect a pot-bound plant.
2 Choose an empty pot that is one size larger than the one that the plant is in.
3 Use crocks to cover the holes in the empty pot and then add a shallow layer of potting compost as shown in figure 11.37 part (a).
4 Spread the fingers of one hand over the surface of the pot-bound plant's soil and turn the pot upside down.
5 Gently knock the rim of the pot against the edge of the bench and remove the plant as shown in figure 11.37 part (b).
6 Remove old crocks and then gently tease out any tightly-bound roots.
7 Place the plant centrally on top of the compost in the new pot as shown in part (c) of the diagram.
8 Fill the space around the plant's roots with potting compost and firm it down as shown in part (d) of the diagram.
9 Place the pot on its drainage saucer and water the plant carefully.
10 Show the pot to your teacher for assessment.

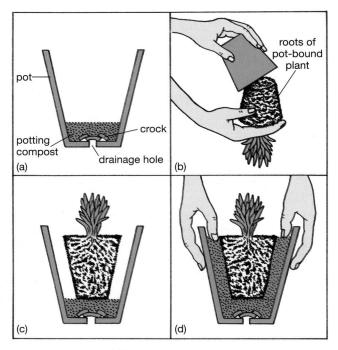

**Figure 11.37** *Potting on*

**Figure 11.38** *Dead-heading a plant*

## Dead-heading

Normally a plant produces a set of flowers which die after making seeds. Under these circumstances, the plant often does not produce any more flowers. However, if the dead flower heads are removed regularly, the plant may produce more flowers.

This removal of dead flower heads to encourage the plant to continue flowering is called **dead-heading** (see figure 11.38). It happens because the plant directs its energy to dormant flower buds instead of to fruit production.

# Pests and diseases

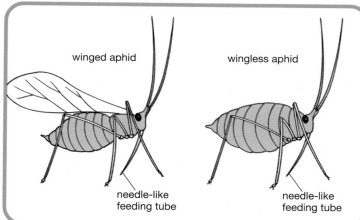

**Figure 11.39**  *Greenfly*

Plants are attacked by many pests and disease-causing organisms.

A **pest** is an animal that causes damage to plants. An example is the **aphid**. The most common type of aphid is the **greenfly** (see figure 11.39), though there are other types such as blackfly. An aphid damages a plant by sucking the plant's sugary sap using mouthparts that take the form of a sharp tube like a syringe needle.

Aphids usually attack the young parts of a plant such as the shoot tips and flower buds. As a result these plant parts do not get their full supply of sugary sap and do not grow properly.

A **disease** is a condition caused by a micro-organism that damages a plant. An example of a disease-causing micro-organism that attacks plants is **grey mould** (scientific name: *Botrytis*). The disease itself is also referred to as grey mould.

Soft-leaved plants are especially vulnerable to this disease. It takes the form of a fuzzy grey fungal growth on the leaves (see figure 11.40). Infected leaves quickly rot and die. The infection soon spreads to the rest of the plant. Stems, flowers and fruit (see figure 11.41) can all be affected. This fungus thrives in damp conditions where ventilation and air circulation are poor.

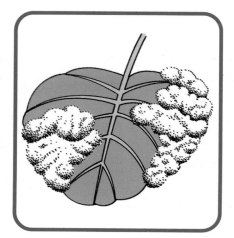

**Figure 11.40**  *Grey mould on a leaf*

**Figure 11.41**  *Grey mould on fruit*

## Methods of controlling pests and diseases

**Figure 11.42** *Chemicals for controlling pests and diseases*

ladybird eating greenfly

**Figure 11.43** *Biological control of greenfly*

### Aphids

#### Chemical control

A **pesticide** is a special chemical that kills insects, mites and other pests. An **insecticide** is a type of pesticide that kills insects. Aphids can be controlled by spraying infected plants with a pesticide (such as Liquid Derris shown in figure 11.42).

Some chemicals work by killing the pest when it comes into contact with the chemical; others work by entering the plant's sap and being sucked up by the insect which is then poisoned and dies. This second type works very well against aphids.

#### Biological control

This means preventing the pest damaging the plant by deliberately introducing one of its natural enemies. This method works well in an enclosed environment such as a greenhouse where, for example, **ladybirds** can be used to keep tomato plants clear of aphids (see figure 11.43).

Some growers disapprove of the use of chemicals to control pests but find biological control outside a greenhouse ineffective. Instead they wash the greenfly off with **soapy water** or simply pick them off the plants and **crush** them to death.

### Grey mould

#### Removal of suitable growing conditions

Grey mould grows well in **damp, stale** air such as that found in a poorly ventilated greenhouse. To control grey mould, all infected and dead or dying plant material should be removed and **burned**. The remaining healthy plants should be well spaced out and given good ventilation. Under these conditions, grey mould cannot survive and the plants will stay healthy.

#### Chemical control

The plants can be sprayed with a type of chemical called a **fungicide** (such as Bordeaux Mixture shown in figure 11.42) which kills fungi such as grey mould.

# Protected cultivation

### Need for protection

In the garden or farm, freshly sown seeds, young seedlings and low-growing crops are all open to damage. This damage can be caused by wind, rain, low air and soil temperatures and, in particular, frost.

However damage can be prevented by protecting the plants. **Protected cultivation** also allows the farmer to sow seeds or plant his crop earlier in the year. Therefore he can harvest his vegetables earlier and get a higher price than he would get later in the year when the market is overflowing with cheap produce.

### Methods of protection

Various materials are used to protect outdoor plants during cultivation.

#### Glass
Glass is often chosen because

- it is good at letting light pass through
- it is permanent
- it is able to retain heat
- it resists being blown over by wind.

Glass is the main component of greenhouses and smaller protective structures such as **cold frames** and some types of **cloche** (see figure 11.44). However, glass is heavy, easily broken and expensive.

#### Plastic
Plastic is often chosen as a protective material because

- it is light in weight making it easy to handle
- it is less easily broken than glass
- it is cheap to buy.

The disadvantage of plastic is that it can reduce the quality of light reaching the plants. In addition, being light in weight, it can be blown away. Despite these potential snags, some rigid forms of plastic make very good cloches (see figure 11.44).

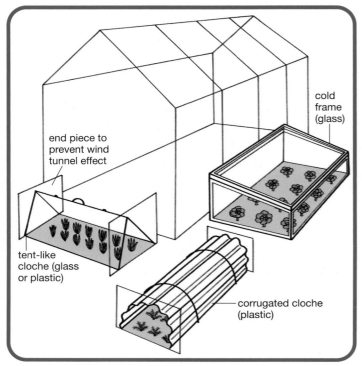

end piece to prevent wind tunnel effect

cold frame (glass)

tent-like cloche (glass or plastic)

corrugated cloche (plastic)

**Figure 11.44** *Cold frame and cloches*

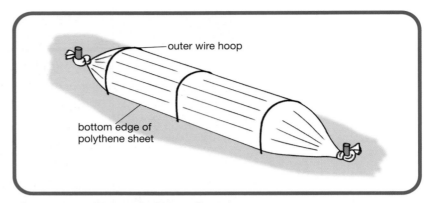

**Figure 11.45**   *Polythene tunnel*

Both glass and plastic cloches keep the soil under them at a higher temperature than that outside. They also protect against frost and wind damage.

## Polytunnels

Where large areas need to be covered cheaply, **polythene tunnels** ('**polytunnels**') made of transparent plastic sheeting stretched over wire hoops are constructed (see figure 11.45). The polythene sheeting is often raised at the sides during the day to ventilate the plants. Low polytunnels give excellent protection to low-growing plants such as strawberries and early carrots.

## Floating fleece

A **floating fleece** is a lightweight sheet of woven man-made fibres (of polypropylene). It is placed over a crop and is gradually lifted up by the plants as they grow (see figure 11.46).

The fleece helps to keep the soil warm and gives protection against frost and many insect pests. At the same time, it allows water, air and light to reach the plants. Some types of vegetable are grown under floating fleeces until they are nearly mature.

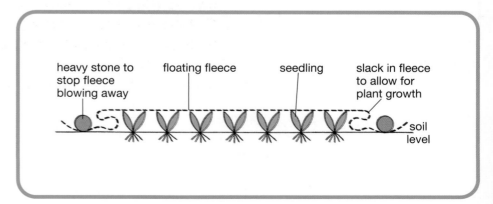

**Figure 11.46**   *Floating fleece*

## Testing your knowledge

1 Decide whether each of the following statements about house plants is true or false and then use T or F to show your choice. Where a statement is false, give the word that should have been used in place of the one in **bold** print. **(6)**
   a) A desert cactus needs **shade** and well-drained soil to grow well.
   b) A fern needs indirect light and **well-drained** soil to thrive.
   c) Foliage plants are grown in people's homes mainly for their decorative **flowers**.
   d) Cacti have fleshy stems that contain **succulent** tissue to give them a store of water.
   e) Most foliage plants like places that are well lit but out of **direct** sunlight.
   f) During the growing season, house plants respond well to regular feeds of **water**.

2 a) (i)  What is meant by the term *pricking out*?
     (ii) Why is it necessary?                                    **(2)**
   b) (i)  What is meant by the term *potting on*?
     (ii) Why is it necessary?                                    **(2)**
   c) (i)  What is meant by the term *dead-heading*?
     (ii) What effect does it have on the plant?                  **(2)**

3 a) (i)  What is an aphid?
     (ii) Give ONE method that a gardener could use to control aphids on a rose bush.                                  **(2)**
   b) (i)  What is grey mould?
     (ii) Give ONE method that you could use to control an attack of grey mould just beginning on an African violet plant.   **(2)**

4 a) Draw a table to show ONE advantage and ONE disadvantage of using
     (i)  glass and
     (ii) plastic to protect young plants.                        **(4)**
   b) (i)  What is a floating fleece?
     (ii) State TWO forms of protection that a floating fleece gives to young plants.                                   **(3)**

## Applying your knowledge

HOUSE PLANT COMPOST: SL SL P P S L

SEED COMPOST: SL SL P P S S L

CUTTING COMPOST: P P S S F L

CACTUS COMPOST: S S S SL G F

AFRICAN VIOLET COMPOST: P P P F L

keys to symbols:
○○○ = large quantity
○○ = medium quantity
○ = small quantity

F = fertiliser
G = gravel
L = lime

P = peat
S = sand
SL = sterilised loam

**Figure 11.47**

**1** Figure 11.47 shows five packs of compost. The ingredients of each type are represented by symbols.

a) Which compost does *not* contain lime? (1)

b) Sort the five composts into two groups:
(i) loamless composts and (ii) loam-based composts. (2)

c) On being watered, which of these composts would hold on to the least water? (1)

d) A sixth type of compost is called multipurpose compost. It was found to contain a little sand, a large quantity of peat and a small amount of fertiliser.

Draw a diagram to show its ingredients represented as symbols. (2)

e) Name a substance that could be used as a substitute for peat in a compost. (1)

| strength of fertiliser to be made up | volume of stock solution to be used (cm³) | volume of water to be used (l) |
|---|---|---|
| full | 100 | 10 |
| half | 50 | 10 |
| quarter | 25 | 10 |

**Table 11.8**

**2** Table 11.8 gives the information needed to make up solutions of a type of liquid fertiliser.

a) A gardener decided that she needed approximately 20 litres of full strength fertiliser. What volume of stock solution should she add to 20 litres of water? (1)

b) A second gardener added 50 cm³ of stock solution to 20 litres of water. What strength of fertiliser did he make up? (1)

c) A third gardener found that she had 25 cm³ of stock solution left and she needed fertiliser of half strength. What volume of water should she use? (1)

| type of flowering plant | time when fertiliser should be used | | | |
|---|---|---|---|---|
| | fertiliser A | fertiliser B | fertiliser C | fertiliser D |
| begonia | up to April | May only | from June on | – |
| dahlia | – | to end of June | from July on | – |
| delphinium | – | up to April | – | from May on |
| fuchsia | up to May | June only | from July on | – |
| gladiolus | – | – | at all times | – |
| rhododendron | – | – | at all times | – |
| rose | up to April | – | from May on | – |
| sweet Pea | up to April | – | – | from May on |

**Table 11.9**

3 Table 11.9 gives information about the use of four different fertilisers on some types of flowering plant.
   a) At which time of the year should rose plants be given fertiliser A? (1)
   b) How many types of plant can be given fertiliser B during June? (1)
   c) How many types of plant can be given fertiliser C during July? (1)
   d) How many different types of fertiliser can be correctly used on fuchsia plants over the six-month period from April to September? (1)
   e) Which fertiliser should be used on dahlia plants in August? (1)
   f) On which TWO types of plant can fertiliser D be used during June? (1)
   g) (i) Fertiliser C has a mineral ratio (N:P:K) of 15:15:30 and fertiliser D has a mineral ratio (N:P:K) of 10:30:20. Draw TWO pie charts to show these ratios.
      (ii) In a 100 kg bag of fertiliser C, how many kilograms would be N? (5)

4 Figure 11.48 shows an experiment set up to investigate the effect of two types of grow bag (X and Y) on two types of tomato plant (Tiny Tim and Outdoor Girl).
   a) Which two numbered parts of the experiment should be compared to find out if grow bag type X or type Y is better for the development of Tiny Tim tomato plants? (1)

b) Which two numbered parts of the experiment should be compared to find out which type of tomato plant grows better on grow bag type Y? (1)
c) Which two numbered parts of the experiment should be compared to find out if two or three plants is the better number to plant in grow bag type X? (1)
d) Name TWO environmental factors that must be kept the same in all six parts of the experiment to make it fair. (2)
e) What could be done to improve the reliability of the results? (1)

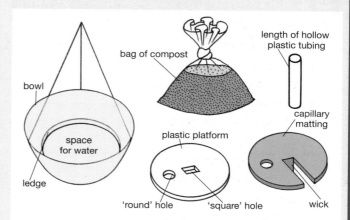

**Figure 11.49**

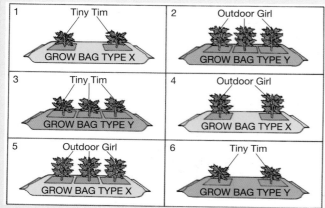

**Figure 11.48**

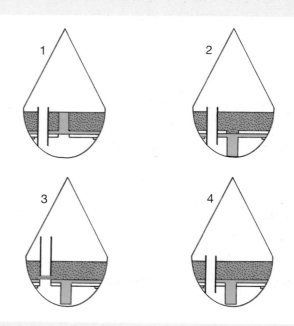

**Figure 11.50**

5 Figure 11.49 shows the equipment needed to set up a type of hanging basket.

Figure 11.50 shows four simple diagrams of the hanging basket set up and ready to receive water and plants.
a) Choose the correct diagram and check your answer with your teacher.
b) Redraw the correct diagram and label all of the parts. (3)

6 The results in table 11.10 were collected using the environmental thermometer shown

| time on 24-hour clock | temperature of soil 4 cm below soil level (°C) |
|---|---|
| 06.00 | 7.0 |
| 09.00 | 7.5 |
| 12.00 | 8.5 |
| 15.00 | 10.0 |
| 18.00 | 10.0 |
| 21.00 | 9.5 |
| 24.00 | 8.5 |
| 03.00 | 8.0 |
| 06.00 | 7.0 |

**Table 11.10**

in figure 11.51. The investigation was carried out in a potato field during the summer.
a) Describe the procedure carried out by the scientists to make use of this type of thermometer. (2)
b) Draw a line graph of the results with time on the x-axis. (4)
c) In general what can be said about the temperature of the soil during the night compared with during the day? (1)
d) By how many degrees Celsius was the temperature of the soil at 21.00 hours

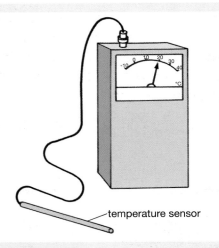

**Figure 11.51**

higher than the temperature of the soil at 09.00 hours? (1)
e) During which three-hour interval did the soil temperature show the greatest increase? (1)
f) Name a piece of equipment that a market gardener could use to protect young lettuce plants at night during cold spring weather. (1)

| state of temp-erature in green-house | response by thermostat | result |
|---|---|---|
| too low | heater's thermostat switches heater on; fan's thermostat switches fan _____ | temperature increases |
| too ____ | _____ thermostat switches heater off; _____ thermostat switches fan _____ | temperature _____ |

**Table 11.11**

7 Table 11.11 refers to the effects brought about by the action of two thermostats in a greenhouse during a day in spring. Copy the table and complete the blanks using the following word bank:

**decreases, fan's, heater's, high, off, on** (6)

8 Match plants W, X, Y and Z shown in figure 11.52 with the four problems given in the following list. (4)
a) Plant has been over-watered.
b) Plant has not been potted on.
c) Plant is being watered wrongly.
d) Plant has been potted on into the wrong size of pot.

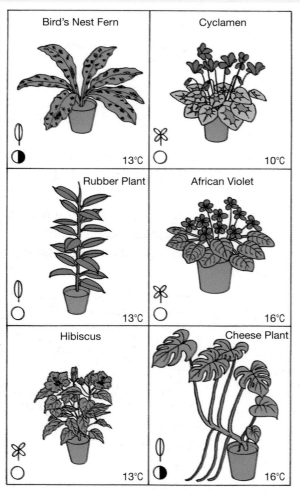

**Figure 11.53**

Key to symbols

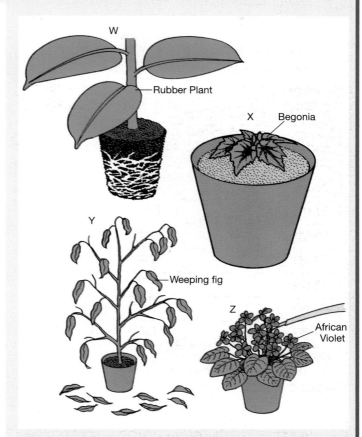

**Figure 11.52**

9 Figure 11.53 gives information about six house plants. Figure 11.54 shows a key that can be used to identify them. The key is incomplete. Copy it down and complete it using the information given in the first diagram. (6)

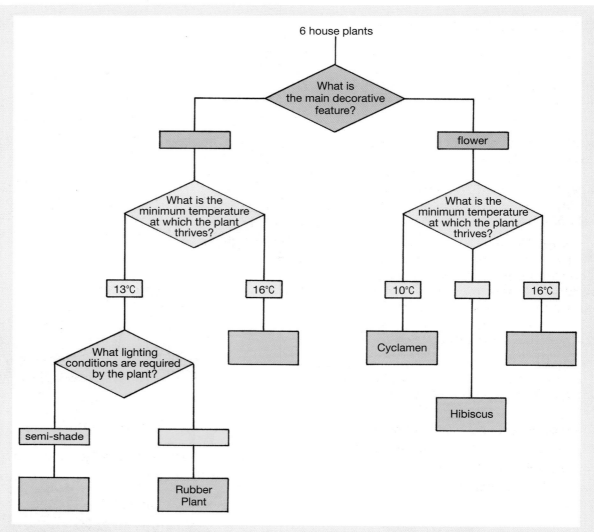

**Figure 11.54**

**10** Read the magazine article on page 299 and then answer the following questions.

a) Make a simple flow chart with SIX steps to show the method to follow when setting up a bottle garden. (6)

b) Why would a desert cactus be an unsuitable choice of plant for a bottle garden? (1)

c) If the hole at the top of the bottle was too small to let your hand in, which of the tools in figure 11.56 would you use to:

(i) pour in gravel, charcoal and compost?

(ii) make the compost level?

(iii) make several planting holes?

(iv) transport a small plant by its roots down into a planting hole? (4)

d) If your bottle garden stayed misted up with condensation all day long, how could you cure this problem without trying to wipe the inside of the glass? (1)

### MAKING A BOTTLE GARDEN

Instead of growing a plant on its own in a pot, a collection of plants can be grown in a large bottle. This works very well with small foliage plants that like humid conditions.

The bottle should be cleaned out with disinfectant, rinsed and allowed to dry. Next, gravel is added for drainage, then a layer of charcoal and lastly a layer of potting compost (see figure 11.55).

If the top of the bottle is too narrow to let a hand in, special tools need to be made (see figure 11.56). Each plant is carefully lowered into a planting hole and tool 2 is used to cover its roots with compost. A small volume of water is added and the bottle's stopper is put in.

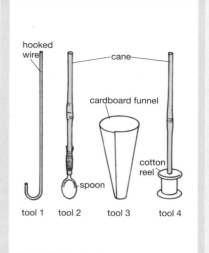

**Figure 11.55**

**Figure 11.56**

If humidity content is correct, drops of condensation will be seen on the upper surfaces of the bottle in the morning and these will disappear during the day. If no drops of water appear at all, your bottle garden needs a little more water.

---

**11** The following simple key identifies some pests and diseases that affect the leaves of house plants and states how to cure them. Use the key to answer the questions that follow it.

**(1)** Are the leaves being damaged by an animal pest?

       Yes . . . . . go to **(2)**
       No . . . . . go to **(15)**

**(2)** Does the animal have jointed legs but no clasper?

       Yes . . . . . go to **(3)**
       No . . . . . go to **(10)**

**(3)** Is the animal green?

       Yes . . . . . go to **(4)**
       No . . . . . go to **(5)**

**(4)** *Your plant has greenfly. Spray it with pesticide.*

**(5)** Is the animal black?

       Yes . . . . . go to **(6)**

       No . . . . . go to **(7)**

**(6)** *Your plant has blackfly. Spray it with pesticide.*

**(7)** Is the animal white?

       Yes . . . . . go to **(8)**
       No . . . . . go to **(9)**

**(8)** *Your plant has whitefly. Spray it with pesticide.*

**(9)** *Seek expert advice.*

**(10)** Does the animal have eight legs?

       Yes . . . . . go to **(11)**
       No . . . . . go to **(12)**

**(11)** *Your plant is under attack by red spider mite. Spray it with pesticide or water.*

**(12)** Does the animal have a long, segmented body with a clasper at one end?

       Yes . . . . . go to **(13)**
       No . . . . . go to **(14)**

**(13)** *Your plant is being eaten by caterpillars. Pick them off and crush them.*

**(14)** *Seek expert advice.*

**(15)** Are the leaves partly covered by a grey, fluffy fungus?

        Yes . . . . . go to **(16)**

        No . . . . . go to **(17)**

**(16)** *Your plant is suffering grey mould disease. Cut off the affected leaves and burn them. Improve ventilation. Apply fungicide as a last resort.*

**(17)** Are there brown spots present on the plant's leaves?

        Yes . . . . . go to **(18)**

        No . . . . . go to **(19)**

**(18)** *Your plant has leaf spot. Cut off the affected leaves and burn them. Keep the plant dry.*

**(19)** Are some of the leaves partly covered with black dust?

        Yes . . . . . go to **(20)**

        No . . . . . go to **(21)**

**(20)** *Your plant has sooty mould. Wipe the leaves clean with a damp cloth.*

**(21)** *Seek expert advice.*

a) (i)   Identify the problem affecting the plant in figure 11.57.

  (ii)  What should be done to cure this problem?      (3)

b) (i)   Identify the pest shown in figure 11.58.

  (ii)  What should be done to this pest when it is found attacking plants?    (3)

c) From the information in the key, state TWO differences between a greenfly and a spider mite.      (2)

**12** Figure 11.59 shows three pieces of gardening equipment in use.

a) What name is given to this type of equipment?      (1)

b) What is it used for?      (1)

c) Explain why the end pieces are important.      (1)

d) Suggest why small gaps are left between the units.      (1)

patch of non-moving black dust

**Figure 11.57**

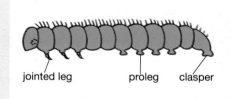

jointed leg      proleg   clasper

**Figure 11.58**

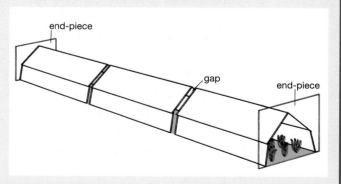

end-piece

gap

end-piece

**Figure 11.59**

## Specimen answer

A specimen answer to *Designing and making a watering system for house plants* (page 277) is shown in figure 11.60.

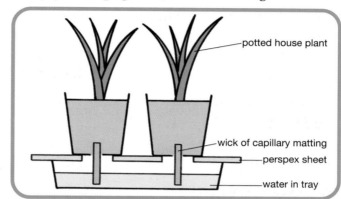

potted house plant

wick of capillary matting

perspex sheet

water in tray

**Figure 11.60**   *Design for watering system*

### What you should know

| | | | |
|---|---|---|---|
| air | fruit | leaves | ratios |
| aphid | fungicides | loam | retentive |
| autovents | glass | matting | root |
| below | grey mould | maximum | sand |
| cocoa | heater | moist | sterilised |
| compost | humid | nutrients | sunlight |
| dead-heading | hygrometer | pesticides | thermometer |
| fan | indirect | plastic | thermostatically |
| fertiliser | irrigation | potting | trace |
| fleece | leaf | pricking | water |

**Table 11.12**   *Word bank for chapter 11*

1   Normal fertile soil is called _____. Man-made growing medium for plants is called _____. If loam is used in a compost, the loam must first be _____ to kill micro-organisms and pests. Loamless compost lacks loam. Instead it contains peat, sand and _____.

2   Peat and substitutes such as _____ shells improve a compost's ability to hold _____. Grit, _____ and perlite improve the _____ content and let excess water drain away easily.

3   Nitrogen (N), phosphorus (P) and potassium (K) are major _____ needed by plants (N for _____, P for _____ and K for flower and _____ growth). Different fertilisers contain different _____ of these minerals. Chemicals needed by plants in tiny amounts are called _____ elements.

4   Some house plants should be watered from above; others from _____. Plants can be left unattended for a week or two by setting up watering systems using capillary _____ or automatic _____. Compost containing water _____ gel does not need to be watered very often because the gel holds much water.

**5** Plants only grow well if given a suitable range of temperature. This can be controlled in a greenhouse using a thermostatically controlled _____. Extremes of temperature can be monitored using a _____ and minimum _____.

**6** Good ventilation in a greenhouse is important to stop the air becoming too _____ (damp) and allowing grey mould to attack the plants. The air can be kept fresh using _____ or a _____-controlled _____. Relative humidity can be measured using a _____.

**7** Cacti are plants that need direct _____ and well-drained soil. Ferns need good indirect light and _____ compost. Foliage plants are grown for their decorative _____; flowering plants for their attractive flowers. In general both of these groups like good _____ sunlight and well-watered compost. However detailed needs vary from one type of plant to another.

**8** The transfer of seedlings from a crowded site to one with more space is called _____ out. The transfer of a potted plant into a larger container is called _____ on. The removal of old flowers to encourage a plant to make more flowers is called _____.

**9** An animal that damages a plant is called a pest. An example is the _____. Such a pest can be killed using chemicals called _____. A disease is a condition caused by a micro-organism. An example that affects plants is _____. Fungal diseases can be treated using chemicals called _____.

**10** Young plants can be protected from extremes of temperature using equipment made of _____ or _____ such as cold frames, cloches and polythene tunnels. Floating _____ can also be used to protect plants from frost.